Math ADVANTAGE

Teacher's Edition

Georgia Test Preparation

CRCT and Stanford 9

Includes Test Taking Tips and math assessment items for
- multiple choice format
- short answer
- extended response

Grade 3

Harcourt Brace & Company
Orlando • Atlanta • Austin • Boston • San Francisco • Chicago • Dallas • New York • Toronto • London
http://www.hbschool.com

Printed in the United States of America

ISBN 0-15-321600-X

2 3 4 5 6 7 8 9 10 073 2003 2002 2001

CONTENTS

Scoring Short Answer Responses

Students can use 3-5 minutes to respond to short-answer test questions. Short-Answer Responses are scored using a rubric. Students can receive partial credit for a partially completed or partially correct answer.

Poor sentence structure, word choice, usage, grammar, and spelling does not affect the scoring of short-answer items, unless communication of ideas is impossible to determine.

Scoring Rubric

Response Level	Criteria
Score 2	**Generally accurate, complete, and clear** ___ All of the parts of the task are successfully completed. ___ There is evidence of clear understanding of key concepts and procedures. ___ Student work shows correct set up and accurate computation.
Score 1	**Partially accurate** ___ Some parts of the task are successfully completed; other parts are attempted and their intents addressed, but they are not completed. ___ Answers for some parts are correct, but partially correct or incorrect for others.
Score 0	**Not accurate, complete, and clear** ___ No part of the task is completed with any success. ___ There is little, if any, evidence that the student understands key concepts and procedures.

Help Students Understand What Scorers Expect

1. Discuss the rubric with students.
2. Have students score their own answer to a practice task, using the rubric.
3. Discuss results. Have students revise their work to improve their scores.

Help students develop proficiency with short-answer questions.

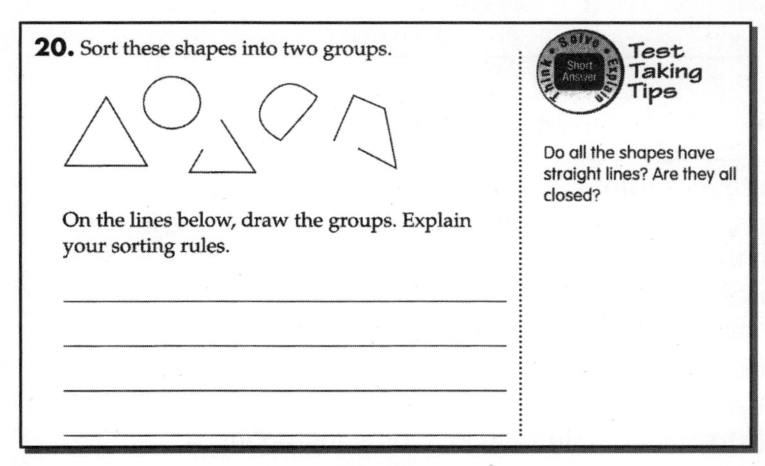

Grade 3

Troubleshooting

Use this discussion to help students answer test items effectively.

"I don't get it!"
Help students read the problem carefully. Then ask, "What do you think you are asked to do?"

"What should I write?"
Have students tell you how they solved the problem. Then have them write their words or use pictures.

"Is this the right answer?"
Have students explain how they know that they have answered the question completely.

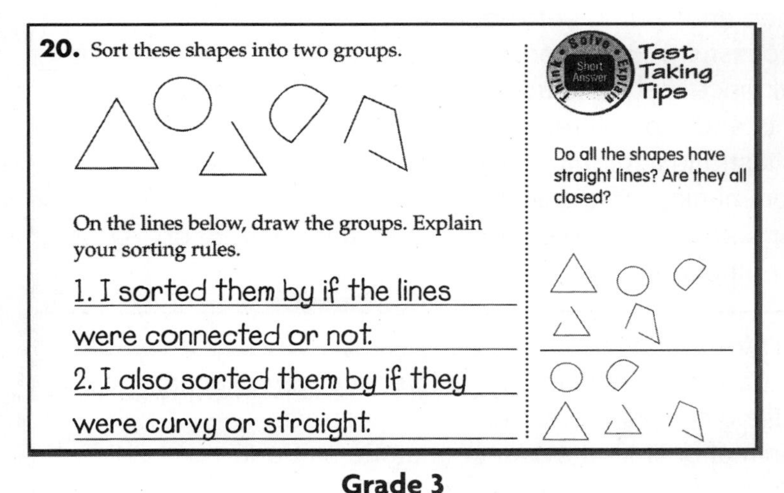

Grade 3

Exemplary response
This student has carefully read the problem. She has shown two sorting rules. She has given a thoughtful explanation.

You may want to make a transparency of this example to share with your students.

Have your students evaluate this response to understand why it is clear and complete.

Scoring Extended Response Items

Performance Task Rubric

Response Level	Criteria
Score 4	**Generally accurate, complete, and clear** ___ All of the parts of the task are successfully completed; the intents of all parts of the task are addressed with appropriate strategies and procedures. ___ There is evidence of clear understanding of key concepts and procedures. ___ Student work and explanations are clear. ___ Additional illustrations or information, if present, enhance communication. ___ Answers for all parts are correct or reasonable.
Score 3	**Generally accurate, with minor flaws** ___ There is evidence that the student has a clear understanding of key concepts and procedures. ___ Student work and explanations are clear. ___ Additional illustrations or information communicate adequately. ___ There are flaws in reasoning and/or in computation, or some parts of the task are not addressed.
Score 2	**Partially accurate, with gaps in understanding and/or execution** ___ Some parts of the task are successfully completed; other parts are attempted and their intents addressed, but they are not successfully completed. ___ There is evidence that the student has partial understanding of key concepts and procedures. ___ Additional illustrations or information, if present, may not enhance communication significantly. ___ Answers for some parts are correct, but partially correct or incorrect for others.
Score 1	**Minimally accurate** ___ A part (or parts) of the task is (are) addressed with minimal success while other parts are omitted or incorrect. ___ There is minimal or limited evidence that the student understands concepts and procedures. ___ Answers to most parts are incorrect.
Score 0	**Not accurate, complete, and clear** ___ No part of the task is completed with any success. ___ Any additional illustrations, if present, do not enhance communications and are irrelevant. ___ Answers to all parts are incorrect.

Help Students Understand What Scorers Expect

1. Discuss the rubric with students.
2. Have students score their own answer to a practice task, using the rubric.
3. Discuss results. Have students revise their work to improve their scores.

Help Students Understand How to Show What They Know

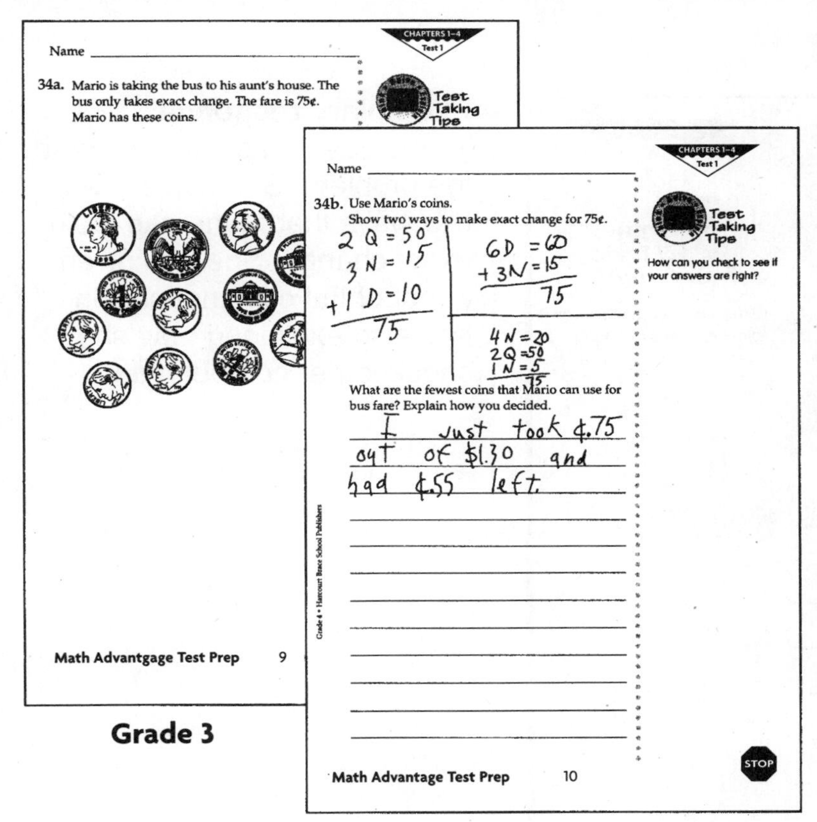

Name ____________

CHAPTERS 1–4 Test 1

34a. Mario is taking the bus to his aunt's house. The bus only takes exact change. The fare is 75¢. Mario has these coins.

Test Taking Tips

Math Advantgage Test Prep 9

Grade 3

Name ____________

CHAPTERS 1–4 Test 1

34b. Use Mario's coins.
Show two ways to make exact change for 75¢.

2 Q = 50
3 N = 15
+ 1 D = 10
75

6D = 60
+ 3N = 15
75

4N = 20
2Q = 50
1N = 5
75

What are the fewest coins that Mario can use for bus fare? Explain how you decided.

I just took $.75 out of $1.30 and had $.55 left.

Test Taking Tips

How can you check to see if your answers are right?

Grade 4 • Harcourt Brace School Publishers

Math Advantage Test Prep 10

STOP

Partially correct response
This student has not read the problem carefully. However, because he did show what he knows about counting money, he will get partial credit for his response.

More practice *reading for information* will help this student read each problem more carefully.

Discussing solution strategies will help him learn to sort through the math knowledge he has to answer the required questions clearly and completely.

Have students evaluate this response to understand how it can be improved.

Ask, "Did the student answer each question?"
Help students see that the student answered the first question but showed more than two ways. The student did not answer the second question. Instead he or she figured out how much change Mario would have left over.

Ask, "What can this student do to improve his or her score?"

The student can restate each part of the problem in his or her own words.
Then the student can check his or her restatement against the problem given and against his or her answer.

Have your students evaluate this response to understand why it is complete.

Daily Practice FCAT

Name ______________________

34b. Use Mario's coins.
Show two ways to make exact change for 75¢.

2Q 4N
2D 1Q
1N 3D

Test Taking Tips

How can you check to see if your answers are right?

What are the fewest coins that Mario can use for bus fare? Explain how you decided.

He could use 5 coins with the ones he has. There are 25 ¢ in a Quarter. + 25+25 = 50 There are 10¢ in a Dime and 10+10 = 20. There is 5¢ in a Nickle. All together 50 + 20 + 5 = 75.

Grade 3

Exemplary response
This student has carefully read the problem. She has shown two ways that Mario can make exact change. She has given a thoughtful explanation. She has also explained how she checked her computation.

Help Students Practice Reviewing and Revising Their Own Work

1. Have a volunteer share a response to a performance task question.
2. Have students discuss the answer.
3. Have students revise their own work to improve their score.

Name ______________________________

Choose the letter of the correct answer.

1. $\begin{array}{r} 4 \\ +5 \\ \hline \end{array}$

A 6 B 7 C 8 D 10 E NOT HERE

Test Taking Tips

Think · Solve · Explain — Multiple Choice

Eliminate choices.

If you solve the problem and don't see your solution listed, mark NOT HERE as the answer.

---

2. $6 - 1 = \underline{\ ?\ }$

F 0 G 3 H 5 J 7 K NOT HERE

3. $\begin{array}{r} 8 \\ -3 \\ \hline \end{array}$

A 2 B 3 C 4 D 5 E NOT HERE

4. $7 + 2 + 5 = \underline{\ ?\ }$

F 13 G 14 H 15 J 16 K NOT HERE

5. $\begin{array}{r} 47 \\ +38 \\ \hline \end{array}$

A 65 B 75 C 85 D 95 E NOT HERE

6. $\begin{array}{r} 60 \\ -37 \\ \hline \end{array}$

F 43 G 33 H 23 J 13 K NOT HERE

7. $\begin{array}{r} 81 \\ -36 \\ \hline \end{array}$

A 15 B 25 C 35 D 45 E NOT HERE

8. $\begin{array}{r} 564 \\ +158 \\ \hline \end{array}$

F 722 G 712 H 622 J 613 K NOT HERE

9. $\begin{array}{r} \$4.18 \\ 1.26 \\ +\ 6.63 \\ \hline \end{array}$

A \$12.07 B \$12.27 C \$12.97 D \$13.06 E NOT HERE

10. Use order in addition to find the missing fact.

$8 + 5 = 13,$

so $\underline{\ ?\ } + \underline{\ ?\ } = \underline{\ ?\ }$

F $5 + 8 = 13$
G $5 + 13 = 18$
H $5 + 8 = 58$
J $8 + 5 = 85$
K NOT HERE

GO ON

Name ______________________________

11. Which subtraction fact is related to this addition fact?

$$7 + 6 = 13$$

A $13 - 6 = 7$
B $7 - 6 = 1$
C $13 + 6 = 19$
D $13 + 7 = 20$

12. Estimate the sum by rounding.

$$\begin{array}{r} 22 \\ +51 \\ \hline \end{array}$$

F 70
G 80
H 90
J 100

13. Linda has 14 red bows and 27 yellow bows for her hair. How many bows does she have in all? Is this number even or odd?

A 31 bows; even
B 31 bows; odd
C 41 bows; even
D 41 bows; odd

14. One day, 17 students were eating lunch. Then 4 students went out to play. How many students were still eating lunch?

F 3 students
G 12 students
H 13 students
J 15 students

15. Kim has 40 animals on her farm. Of these, 17 are cows. The rest are sheep. How many sheep are on Kim's farm?

A 3 sheep
B 13 sheep
C 17 sheep
D 23 sheep

16. James had some money. His father gave him $2.00 more. James bought a game for $4.00 and a toy car for $2.50. He now has $1.50 left. How much money did James have to begin with?

F $1.00
G $5.00
H $6.00
J $8.25

17. Kim's grandparents live 400 miles away. Her aunt and uncle live 263 miles closer. How many miles away do her aunt and uncle live?

A 37 miles
B 137 miles
C 139 miles
D 317 miles

18. During one month, 185 tickets were sold at Game A, 234 tickets were sold at Game B, and 168 tickets were sold at Game C. How many tickets were sold at all three games?

F 577 tickets
G 587 tickets
H 684 tickets
J 5,187 tickets

19.
$$\begin{array}{r} 821 \\ -379 \\ \hline \end{array}$$

A 334
B 432
C 442
D 552

Name ________________________________

Jaime started out with 16 white socks. Later, he had only 9 white socks. How many socks got lost?

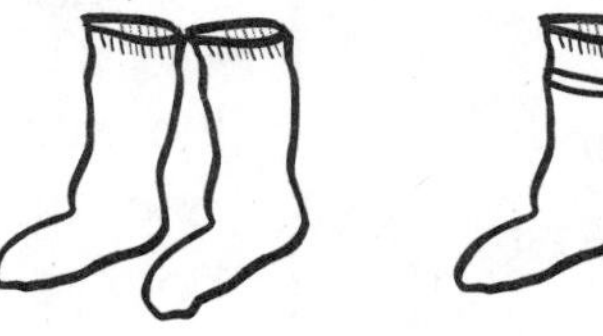

On the lines below, explain how you figured out the answer.

____

____

____

____

What do you know about fact families that will help you solve the problem?

Tara has 15 crayons. Her friend Ellen has none. Tara wants to share as equally as possible, but she does not want to break any of the crayons.

Find two ways that Tara can share her crayons with Ellen. Write number sentences to show how you solved the problem.

____

____

What do you know about fact families that will help you solve this problem?

Name ______________________________

Marie has 4 boxes of holiday decorations. The first box holds 18 items. The second one holds 22 items. The third box holds 9 items. The fourth one holds 12 items.

About how many holiday decorations does Marie have? Explain the strategy you used to make your estimate.

____

____

____

____

____

____

How can rounding up and rounding down help you solve the problem?

23 Ms. Sampson gave out 105 pencils from this box.

About how many pencils are left in the box?

____

Explain how you found your answer.

____

____

____

Test Taking Tips

Are you looking for an exact answer or an estimate?

Name ______________________________

24 Keith's class of 29 students was putting on a play. There were acting roles for only 7 students. How many students will not have acting roles?

On the lines below, show your work and explain your thinking.

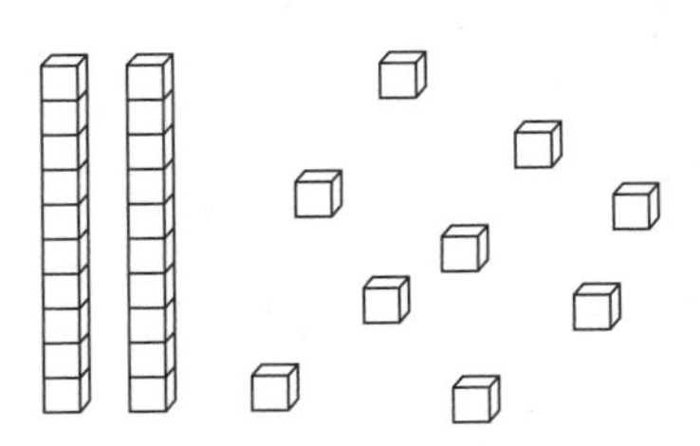

____

____

How can you use base-ten blocks to model the problem?

25 How much faster can the zebra run than the rabbit?

Animal	Top speed in kilometers per hour
Cheetah	112
Lion	80
Elk	72
Zebra	64
Rabbit	56
Reindeer	51

Show how you solved the problem.

____

____

____

What information do you need from the chart before you can solve this problem?

GO ON

Name ______________________________

26 On Monday and Tuesday, the vet treated 53 dogs in all. On Monday, she treated 29 dogs. How many dogs did she treat on Tuesday?

On the lines below, explain the method you used to figure out the problem.

____

____

____

____

Solve • Explain • Think — Short Answer

Test Taking Tips

How can you restate the problem?

27 Lisa left the library and walked to the park. From there, she walked to the post office. How far did she walk?

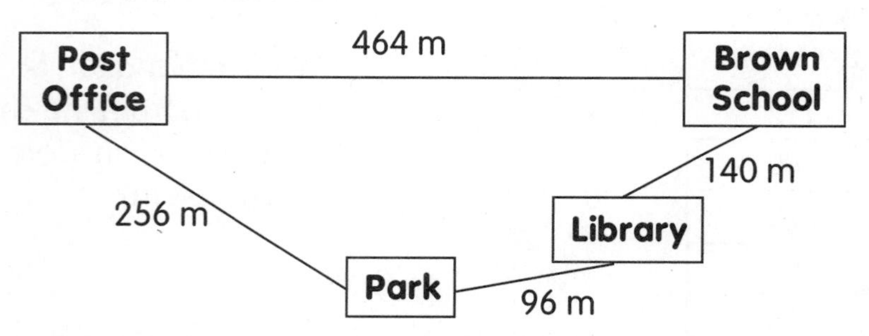

On the lines below, show your work and explain your thinking.

____

____

____

Solve • Explain • Think — Short Answer

Test Taking Tips

What information from the map is useful in solving this problem?

Name ______________________________

28 a. Paul needs 20 pies for a party. He has baked 8 of them. His two aunts have baked 2 pies each. His mother has baked 4. Paul hopes his father will bake the rest. How many pies will his father have to bake?

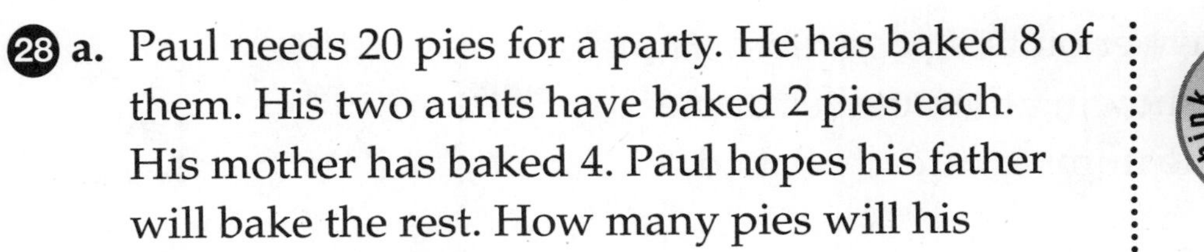

Test Taking Tips

Think • Solve • Explain — Long Answer

How many pies have already been baked?

GO ON

Name ____________________

28 b. Write your answer on this page. Draw a picture or a diagram showing the number of pies needed and the number already baked.

Explain how you know your answer is right.

____

____

____

____

____

____

____

____

____

Think • Solve • Explain — Long Answer

Test Taking Tips

How can drawing a picture help you solve the problem?

Name ______________________________

a. Pablo's father wanted to treat his children to a tour of St. Augustine, Florida. He read the following sign at the tourist center:

Test Taking Tips

What information from the chart do you need?

Tours of St. Augustine, Florida

Train Tickets
Adult $7.00 Child $2.00

Horse and Carriage Rides
Adult $7.00 Child $3.00

Pablo's father paid $20 for horse and carriage tickets.

How many adults and how many children went on the ride?

GO ON

Name ______________________________

29 b. Write your answer on this page. Explain your reasoning.

How can you check your answer?

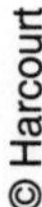

Name ______________________________

Choose the letter of the correct answer.

For questions 1–2, tell what time it is.

1

A 6 minutes after two
B 6 minutes after three
C 23 minutes after six
D NOT HERE

Test Taking Tips

Get the information you need.
Which part tells the hour on a digital clock?

2

F 7:20 **H** 8:24
G 7:25 **J** 8:30

3 Tell how many minute marks the minute hand has moved from 12. Count by fives.

A 7 minutes
B 30 minutes
C 35 minutes
D 40 minutes

4 The hour hand on Bob's watch is a little past five. The minute hand is pointing to the 2. What time is it?

F 5:10 **H** 6:02
G 5:20 **J** 6:10

5 Tim is meeting Richard at the park at 4:15. Where will the minute hand on the clock be then?

A on the 11 **C** on the 6
B on the 9 **D** on the 3

6 Jenny's mother asked her to help cook dinner. How much time will it take Jenny to help cook dinner?

F about 3 minutes
G about 30 minutes
H about 30 hours
J about 300 hours

For questions 7–8, use the schedule.

SCHOOL FUN WEEK		
Activity	**Day**	**Time**
3-Legged Race	Monday	9:00–9:30
Tug-of-War	Tuesday	10:00–10:30
100-m Race	Wednesday	8:30–?
Balloon Toss	Thursday	11:00–11:30
200-m Race	Friday	11:30–12:00

7 The 100-meter race lasts for 1 hour and 30 minutes. At what time does it end?

A 8:30 **C** 10:00
B 9:00 **D** 10:30

Name ________________________________

8. Ice cream is half price at the same time that the Tug-of-War takes place. When is ice cream half price?

F Monday 9:00 – 9:30
G Tuesday 10:00 – 10:30
H Thursday 11:00 – 11:30
J Friday 11:30 – 12:00

9. When the digital clock says 8:00, what time is it?

A eight o'clock
B 5 minutes after eight
C nine o'clock
D NOT HERE

10. Count the money.

F $1.21
G $1.26
H $1.31
J $1.36

11.
$$\begin{array}{r} \$32.67 \\ +\ 14.82 \\ \hline \end{array}$$

A $46.39
B $46.49
C $47.39
D $47.49

12.
$$\begin{array}{r} \$7.65 \\ -\ 0.82 \\ \hline \end{array}$$

F $5.82
G $6.73
H $6.83
J NOT HERE

13. Pam spends $6.25 for a theme park ticket, $1.50 for chips, and $2.50 for a puzzle. How much money does she spend in all?

A $10.25
B $9.25
C $9.00
D NOT HERE

For questions 14–15, use the calendar.

June

Sun	Mon	Tue	Wed	Thu	Fri	Sat
			1	2	3	4
5	6	7	8	9	10	11
12	13	14	15	16	17	18
19	20	21	22	23	24	25
26	27	28	29	30		

14. What is the date three weeks before June 23?

F June 1
G June 2
H June 3
J June 30

15. Sarah kept a record of the rainfall for each day starting on June 12. She kept a record for the next 11 days. What was the last day she recorded the rainfall?

A June 23
B June 26
C June 28
D June 30

GO ON

Name ______________________________

Sean's schedule for his guitar lessons was chewed up by the dog.

Guitar Lessons	
Monday	12:30
Tuesday	12:45
Wednesday	
	1:15
	1:30
Saturday	

Look for the pattern. Fill in the missing days and times.

On the lines below, explain the pattern.

____

____

How can counting by 15's help you find a pattern?

On Monday, Gillian saw the sunrise at 6:30 A.M. That evening, she saw the sunset at 6:48 P.M.

How many hours and minutes of daylight were there on Monday?

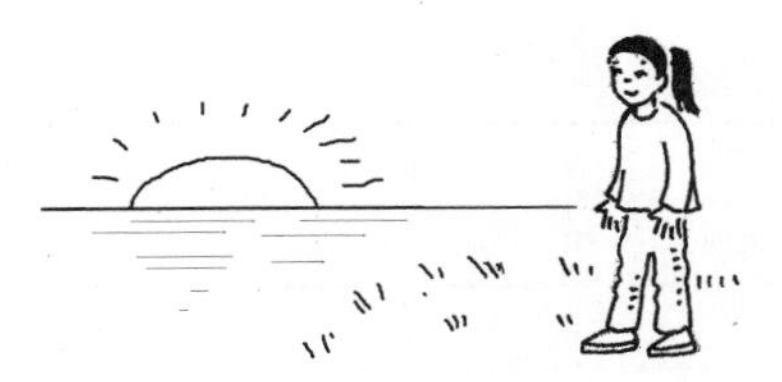

____

On the lines below, explain how you solved the problem.

____

____

____

What are you trying to find out? Restate the problem in your own words.

GO ON

Name ______________________________

18 Cheryl has been waiting to use a treadmill, but all the treadmills are in use. It is now 4:17.

This chart is posted near the treadmills. How much longer will Cheryl have to wait before she can begin her exercise?

Treadmill Sign-up Sheet
Time Limit: 30 minutes

Name	Start Time
Dwaine	4:08
Suzanne	3:50
Oscar	4:15
Loretta	3:59

Explain how you figured out the answer.

____

____

How can finding each stop time help you solve the problem?

19 Elizabeth bought a baseball bat for $17.23. She gave the cashier $20.00. How much change should Elizabeth get back? What coins and bills could the cashier use to give this change to Elizabeth?

On the lines below, explain how you solved the problem.

____

____

How can acting it out help you solve this problem?

Name ______________________________

The chart shows the ways Carla can earn money.

Carla wants to buy a ticket to go to the circus. Tickets cost $8.75. How many times will she have to wash the dishes to earn enough for a circus ticket?

Ways to Earn Money	
Sweep kitchen floor	$1.00
Mow lawn	$2.35
Wash dishes	$1.50
Walk the dog	$2.00

On the lines below, explain how you solved the problem.

____

____

Solve · Think · Explain — Short Answer

Test Taking Tips

What operation can you use to solve the problem?

21 Marco was looking for some camping gear. One store advertised the following prices:

Camping Gear	
Backpack	$9.67
Flashlight	$8.85
Food Cooler	$6.69
Air Mattress	$7.98

Which item is least expensive?

Which item is most expensive?

On the lines below, explain how you know.

____

____

____

Test Taking Tips

How can ordering prices from most to least help you solve this problem?

Name ______________________________

 Julio chose a sweater for $19.95, pants for $14.96, and 2 pairs of socks for $4.95 each. ESTIMATE how much money he will need to pay for them.

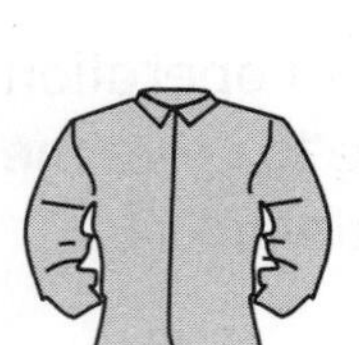

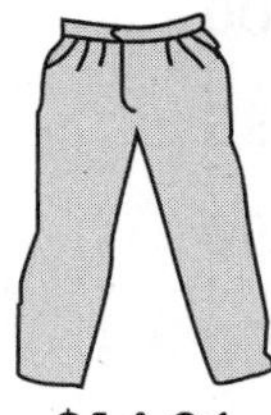

$14.96

$4.95

$4.95

On the lines below, explain the steps you followed to find the answer.

____

____

____

Test Taking Tips

How can rounding to the nearest dollar help you solve the problem?

 Anthony went to the bookstore and picked out these three books.

Anthony has $20. Does he have enough money to buy the three books?

Use words and numbers to explain your thinking.

____

____

____

Test Taking Tips

What are you trying to find out? Restate the problem in your own words.

GO ON

Name ______________________________

a. Kristin looked at the clock at 5:30. She realized that her birthday party would be over in 30 minutes. It started three hours ago.

How can drawing a picture help you solve the problem?

Draw hands on the clock faces. On one, show the time the party started. On another, show the time Kristin looked at the clock. On the third, show the time the party will be over.

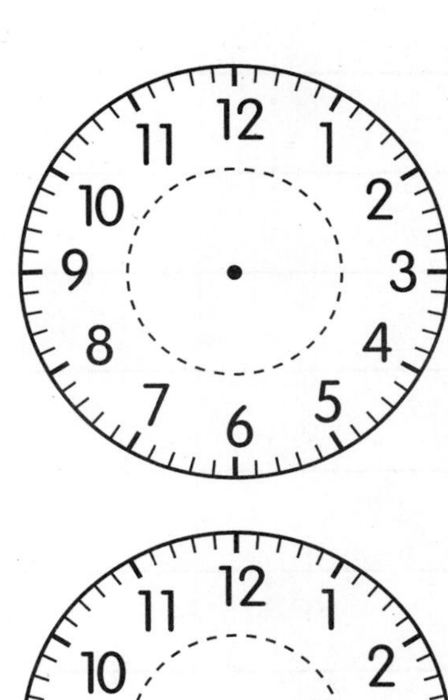

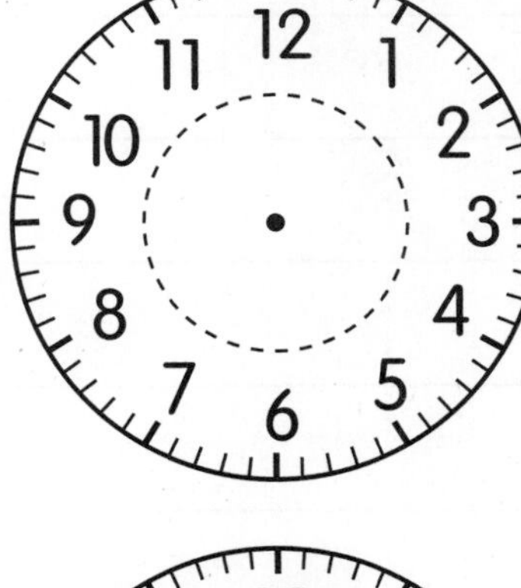

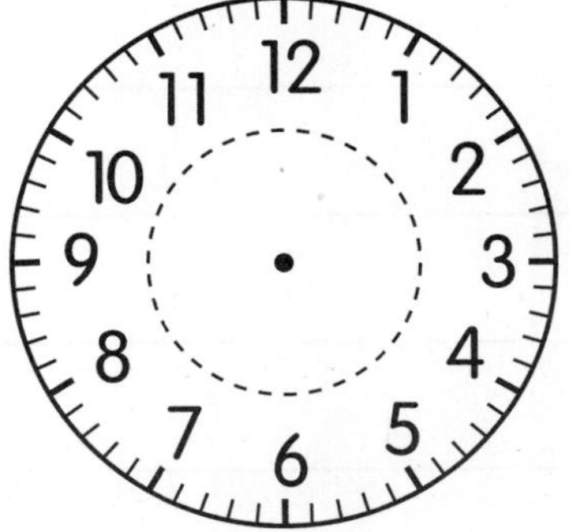

24 **b.** At what time did the party start? ____:____

At what time did Kristin look at the clock? ____:____

At what time did the party end? ____:____

Explain how you checked your answer.

____

____

____

____

____

____

____

____

____

____

____

____

____

____

____

____

____

How can rereading the question help you check your answer?

GO ON

Name ____

a. Brian made tally marks on a chart as he counted the change in his pockets.

Half Dollars	Quarters	Dimes	Nickels	Pennies
𝍷𝍷𝍷	𝍸	𝍸 𝍷𝍷	𝍸 𝍸 𝍷	𝍷𝍷𝍷

Does Brian have enough money to buy a baseball cap for $4.95? If so, how much will he have left over? If not, how much more money does he need?

How can making a list help you find the answer?

Name ________________________________

25 b. Write your answers on this page. Show your work and explain your reasoning.

Be sure that your explanation is clear and complete.

STOP

Name ________________________________

Choose the letter of the correct answer.

1 How many [tens rod] are there in [hundreds flat]?

A 10 C 70
B 50 D 100

Solve · Explain · Think — Multiple Choice

Test Taking Tips

Understand the problem.
How could you use skip-counting to find the answer?

2 How many nickels are equal to one dime?

F 1 H 10
G 2 J 25

For questions 3–4, use the calendar.

April

Sun	Mon	Tue	Wed	Thu	Fri	Sat
		1	2	3	4	5
6	7	8	9	10	11	12
13	14	15	16	17	18	19
20	21	22	23	24	25	26
27	28	29	30			

3 What is the date of the fifth Tuesday?

A April 8 C April 23
B April 15 D April 29

4 How many Fridays are in the month?

F 2 Fridays H 4 Fridays
G 3 Fridays J 5 Fridays

5 What is the value of the 5 digit in 562?

A 5 C 500
B 50 D 5,000

For questions 6–7, use patterns of tens to find the sum or difference.

6 $37 + 30 = \underline{?}$

F 40 H 57
G 47 J 67

7 $95 - 40 = \underline{?}$

A 25 C 40
B 35 D 55

8 Jess is holding 5 coins in her hand. The coins are worth 47 cents. Which set of coins is Jess holding?

F 2 quarters, 2 nickels, 1 penny
G 1 quarter, 2 dimes, 2 pennies
H 1 quarter, 2 dimes, 1 penny
J NOT HERE

For questions 9–10, use patterns of hundreds or thousands to find the sum or difference.

9 $734 - 400$

A 334 C 534
B 434 D 637

Name ______________________________

10 4,503 + 2,000

F 2,503 H 6,503
G 4,505 J 7,503

11 What number is six thousand nine hundred forty-six?

A 6,694 C 9,264
B 6,946 D 9,462

12 What is the number?

50,000 + 8,000 + 100 + 20 + 7

F 51,127 H 58,712
G 58,127 J NOT HERE

For questions 13–14, use the table.

Classes in Oak Hill School are collecting soup labels to buy a jungle gym for their school.

SOUP LABEL COLLECTION	
Room Number	Number of Soup Labels
12	8,000
13	12,000
14	16,000
15	6,500
16	18,000

13 Room helpers have picked up labels in rooms that have collected more than 15,000 labels. Which two rooms have the room helpers picked up labels from?

A Rooms 13 and 14
B Rooms 13 and 16
C Rooms 14 and 16
D Rooms 15 and 16

14 It will take about 1 hour to count 1,000 labels. About how long will it take to count the labels in Room 13?

F about 1 hour
G about 12 hours
H about 120 hours
J about 1,200 hours

15 Which answer shows the numbers in order from least to greatest?

528, 544, 488

A 544, 528, 488
B 528, 544, 488
C 488, 544, 528
D 488, 528, 544

16 Use the number line. Which two tens is the number 48 between?

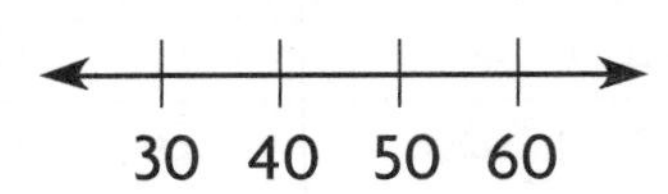

F 20–30 H 40–50
G 30–40 J NOT HERE

17 Round the number 780 to the nearest hundred.

A 500 C 700
B 600 D 800

18 Compare the numbers. Choose $<$, $>$, or $=$.

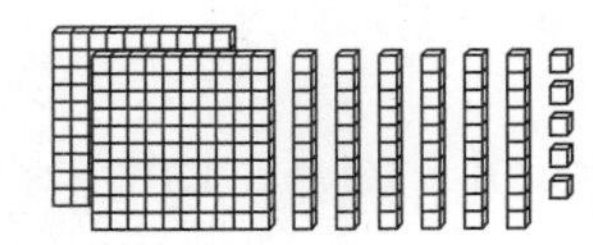 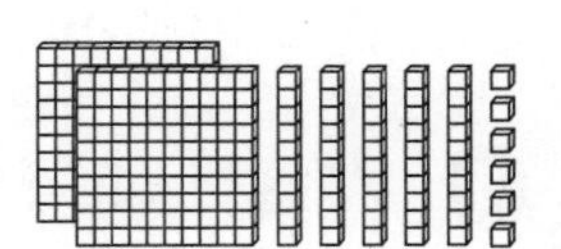

265 ● 256

F $<$ G $=$ H $>$

Name ______________________________

What is the value of each digit in 405?

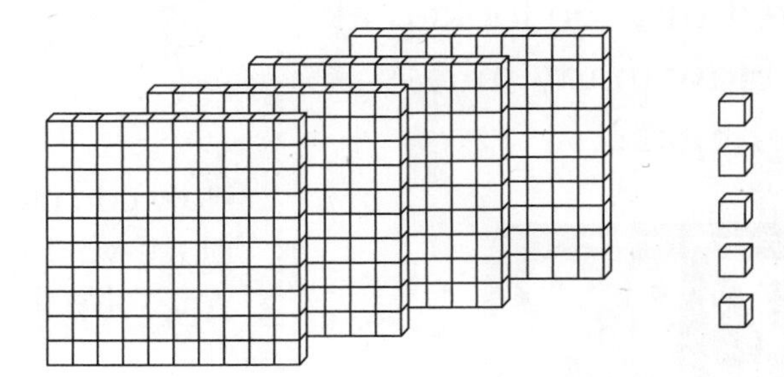

On the lines below, explain how you know.

____

____

____

____

Solve • Short Answer • Think • Explain

Test Taking Tips

How can you use place value to answer the question?

Madison School took a survey to find out how students get to school. How many third graders do NOT walk to school? Show how you solved the problem. Use numbers and words.

How Third Graders Get to School	
Ride the bus	18
Walk	12
Ride bikes	14
Ride in a car	11

____

____

____

What information in the chart can help you solve the problem?

Name ____

21 The odometer in Mr. Langley's car read 6,592 when he bought gas. The next time he looked at the odometer, it read 6,792. How many miles had he driven since he bought gas?

006,792

On the lines below, explain how you decided.

____

____

____

How can using a place-value chart help you solve this problem?

22 Beth played soccer for 35 hours during the month of January. Lori played soccer for 53 hours. Who played more? How do you know?

On the lines below, explain how you decided.

____

____

____

____

What do you know about place value that will help you solve this problem?

GO ON

Name ____________________

23 The diagram shows how the first three houses on Clara's side of the block are numbered.

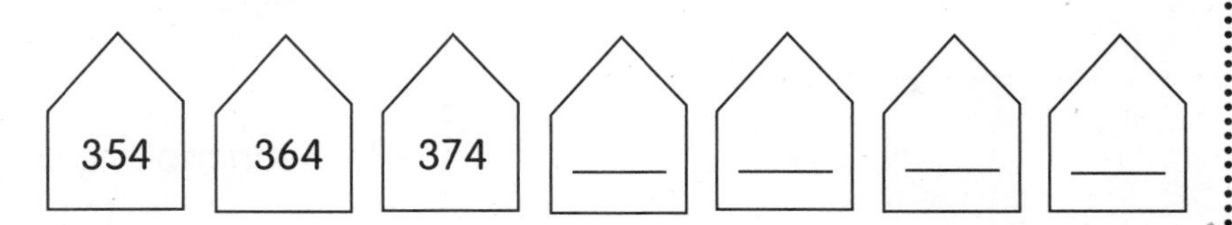

What are the house numbers for the next four houses?

On the lines below, explain how you decided on the numbers for the next four houses.

____

____

____

How can finding a pattern help you solve the problem?

24 Harry has to order some more horse feed, to replace what he sold in the past week. The list shows what he sold.

Horse Feed Sold

Monday	57 pounds
Tuesday	43 pounds
Wednesday	68 pounds
Thursday	32 pounds
Friday	47 pounds

On the lines below, tell if an estimate of 250 pounds is reasonable. Explain your reasoning.

____

____

____

When do you round up? When do you round down?

Name ______________________________

Marvin's Food Mart is 157 miles from the town of Swanee, 261 miles from Burnic, and 176 miles from Sterling.

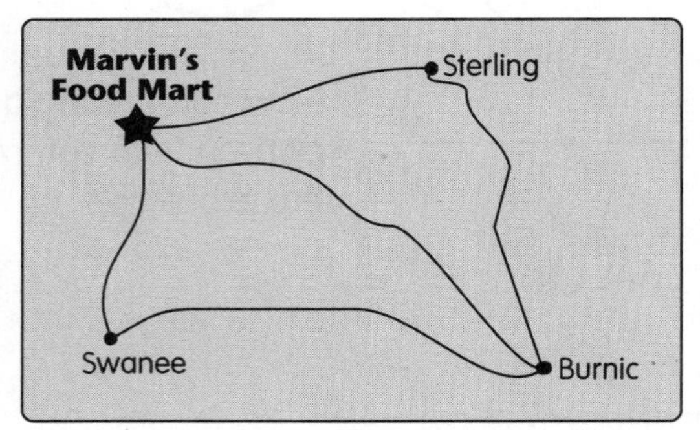

Map

Which town is closest to Marvin's Food Mart? Which one is farthest from the Mart?

On the lines below, explain how you solved the problem.

____

____

____

How can making a list help you solve the problem?

The person in charge of the cafeteria kept track of the lunches sold in one day. She made this list:

Monday	700
Tuesday	767
Wednesday	576
Thursday	764
Friday	746

What was the largest number of lunches sold in one day? Explain.

____

____

Test Taking Tips

How can ordering the numbers help you solve the problem?

Name ______________________________________

27 **a.** Carmella started making a table to show prizes for the fair. She had to stop before she was finished.

Look for a pattern.

Help Carmella finish the table.

Number of Tickets Won	Prize Value
8	$2.00
10	$2.50
12	$3.00
14	
	$4.00
18	
20	
	$5.50

How can finding a number pattern in each column help you complete the table?

GO ON

Name ______________________________

27 **b.** Explain how you decided to complete your table.

____

____

____

____

____

____

____

____

____

____

____

____

____

____

____

____

____

____

____

How can you check that your solution makes sense?

Name ________________________________

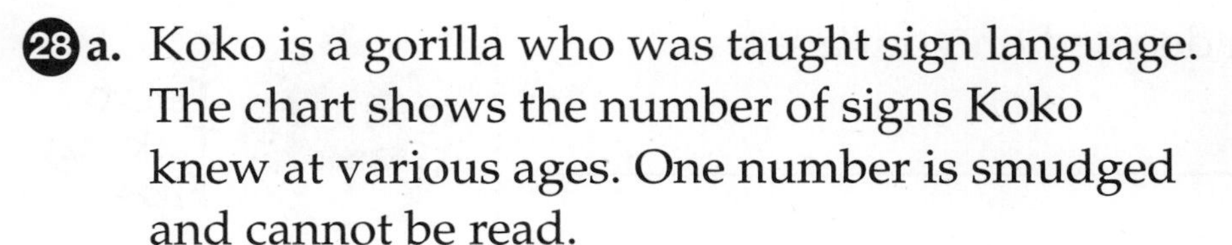

28 a. Koko is a gorilla who was taught sign language. The chart shows the number of signs Koko knew at various ages. One number is smudged and cannot be read.

Age in Months	Number of Signs
42	111
44	135
46	[illegible]
48	182
50	199

Look for a pattern in the chart.

What is a reasonable number of signs that Koko might know at 46 months?

____

____

____

____

____

____

____

____

____

____

____

Think • Solve • Explain — Long Answer

Test Taking Tips

About how many signs does Koko learn every 2 months?

GO ON

Name ______________________________

28 b. Explain how you decided on your answer.

How can you check your work?

STOP

Name ______________________________

Choose the letter of the correct answer.

1 $1 \times 8 = \underline{\ ?\ }$

A 0
B 1
C 8
D 18
E NOT HERE

Test Taking Tips

Decide on a plan.

Use what you know about multiplying by 1 to solve the problem.

---

2 $\begin{array}{r} 9 \\ \times 3 \\ \hline \end{array}$

F 12
G 18
H 24
J 27
K NOT HERE

3 Todd sleeps for 8 hours every night. How many hours will he sleep in 4 nights?

A 12
B 24
C 28
D 32
E NOT HERE

4 A cabinet had 5 shelves with 6 glasses on each shelf. How many glasses were there in all?

F 11 glasses
G 20 glasses
H 30 glasses
J 40 glasses
K NOT HERE

5 $6 \times 3 = \underline{\ ?\ }$

A 24
B 18
C 16
D 12
E NOT HERE

6 $\begin{array}{r} 7 \\ \times 8 \\ \hline \end{array}$

F 24
G 32
H 40
J 56
K NOT HERE

7 $36 \div 9 = \underline{\ ?\ }$

A 3
B 4
C 5
D 6
E NOT HERE

8 $28 \div 7 = \underline{\ ?\ }$

F 4
G 5
H 6
J 7
K NOT HERE

9 $45 \div 5 = \underline{\ ?\ }$

A 6
B 7
C 8
D 9
E NOT HERE

10 $21 \div 7 = \underline{\ ?\ }$

F 0
G 1
H 2
J 3
K NOT HERE

11 Bob put cookies on a tray for a picnic. He made 3 rows with 8 cookies in each row. How many cookies were there in all?

A 24
B 36
C 48
D 54
E NOT HERE

GO ON

Name ______________________

12 Which two smaller arrays can be used to find the product 6×7?

7   7
3   3

F 3×7 and 3×7
G 3×7 and 2×7
H 4×7 and 3×7
J 5×7 and 2×7

13 Julie put class pictures in an album. She put 9 pictures in each row. There were 3 rows. How many pictures does she have?

A 12 pictures
B 24 pictures
C 27 pictures
D 36 pictures

14 Choose the division sentence shown by the repeated subtraction.

$$\begin{array}{r}16\\-\ 4\\\hline 12\end{array}\qquad\begin{array}{r}12\\-\ 4\\\hline 8\end{array}\qquad\begin{array}{r}8\\-4\\\hline 4\end{array}\qquad\begin{array}{r}4\\-4\\\hline 0\end{array}$$

F $16 \div 2 = 8$ **H** $12 \div 4 = 3$
G $16 \div 4 = 4$ **J** $8 \div 4 = 2$

15 For a party, Mrs. Holt pours 6 ounces of juice into each glass. She has a 36-ounce pitcher of juice and an 18-ounce pitcher of juice. How many glasses can she fill?

A 6 glasses **C** 8 glasses
B 7 glasses **D** 9 glasses

16 Choose the number sentence that solves the problem.

Students in a craft class are making clowns. They need 4 pieces of yarn for each clown. How many pieces of yarn do they need for 6 clowns?

F $6 \times 4 = 24$
G $6 + 4 = 10$
H $6 \div 2 = 3$
J $6 - 4 = 2$

17 Use the number line.

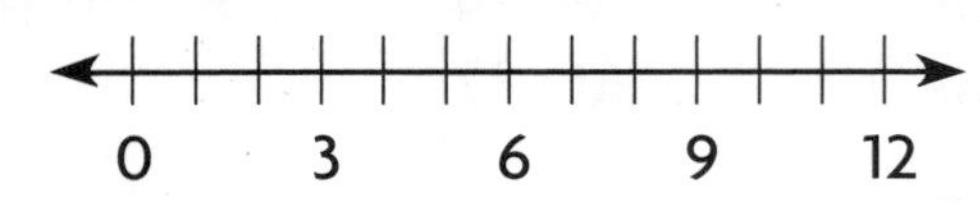

$6 \times 2 = \underline{\ ?\ }$

A 21 **C** 16
B 18 **D** 12

18 What is the missing number for each number sentence?
$5 \times \underline{\ ?\ } = 35$ $\qquad 35 \div 5 = \underline{\ ?\ }$

F 6 **H** 8
G 7 **J** 9

19 ○ ○ ○ ○

$4 \div 4 = \underline{\ ?\ }$

A 0 **C** 2
B 1 **D** 5

20 Beth has 16 small dolls. She placed an equal number of dolls in each of 4 piles. How many dolls were in each pile?

F 2 dolls **H** 6 dolls
G 4 dolls **J** 8 dolls

GO ON

Name ____

21 Marly can write her name 4 times in 1 minute. How many times can she write it in 4 minutes?

Marly

Explain how you solved the problem.

____

____

____

Think • Solve • Explain — Short Answer

Test Taking Tips

How can using a multiplication table help you solve this problem?

22 Helena wanted to buy 4 muffins for each person in her family. There are 6 people in her family. How many muffins should she buy?

Use words and pictures to explain your answer.

____

____

____

Think • Solve • Explain — Short Answer

Test Taking Tips

How can drawing a picture help solve the problem?

What operation can you use to solve the problem?

GO ON

Name ______________________________

23 The elevator in the apartment building takes 7 seconds to move from one floor to the next.

How long will it take to travel from Floor 2 to Floor 7?

Explain how you found your answer.

____

____

____

What do you need to figure out first?

---

24 Jamal noticed a pattern in these numbers.

6, 12, 18, 24, 30,…

What would be the next three numbers in the pattern?

____

Describe the pattern that helps you predict the next numbers.

____

____

____

____

How is each number related to the number that comes before it?

Name ______________________________

25 There are 4 basketball teams at Hightower School. Each team has 9 players. How many basketball players are there in all? Explain how you know your answer is correct.

____

____

____

Think · Solve · Explain — Short Answer

Test Taking Tips

What operation can you use to solve the problem?

26 William poured nickels into a coin-counting machine. The machine said he had 45¢.

How many nickels did William pour into the machine?

____

Explain how you know.

____

____

____

Think · Solve · Explain — Short Answer

Test Taking Tips

How could counting by fives help you solve the problem?

Could you use an operation?

GO ON

Name ______________________________

27 Selena is arranging 72 cookies on a plate. Each row has 8 cookies. How many rows of cookies can she make?

Think · Solve · Explain
Short Answer
Test Taking Tips

How can making an array help you solve this problem?

On the lines below, explain the method you used to figure out the problem.

____

____

____

____

28 What calculator key is missing?

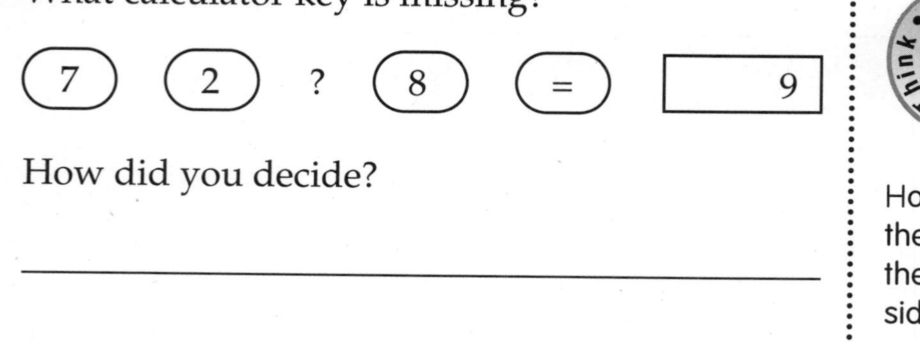

How did you decide?

____

____

____

How does the answer in the display compare to the numbers on the left side of the equation?

Name ______________________________

a. Keesha runs a day-care center. She needs to buy equipment for four play groups. Each group needs a baseball, a soccer ball, and a kite. About how much money does she need for each play group?

Baseball	Soccer Ball	Kite
$2.15	$1.99	$2.75

How can rounding to the nearest dollar help you find the answers?

GO ON

Name ______________________________

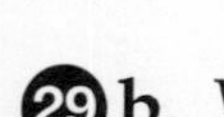

29 b. Write your answers on this page. Show your work and explain your reasoning.

How can using a multiplication table help?

____

____

____

____

____

____

____

____

____

____

____

____

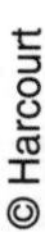

GO ON

Name ______________________________

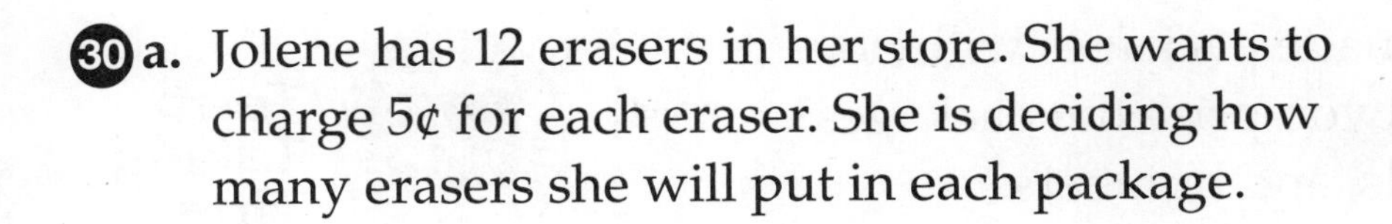

30 a. Jolene has 12 erasers in her store. She wants to charge 5¢ for each eraser. She is deciding how many erasers she will put in each package.

How can you use the picture to help solve the problem?

Help Jolene find three different ways to divide the 12 erasers into packages, ending up with no leftovers. Record your ideas below.

Number of Packages	Number of Erasers in One Package	Price per Package
		¢
		¢
		¢

GO ON

Name ______________________________

30 b. Explain how you made packages with no leftovers. How did you price the packages? Use pictures, words, and numbers.

How can you make a list to solve the problem?

STOP

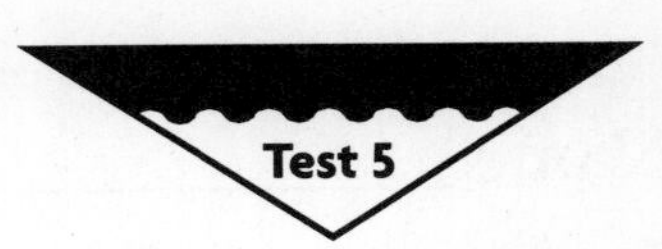

Name ______________________________

Choose the letter of the correct answer.
For questions 1–2, use the pictograph.

OUR FAVORITE WHEEL RIDES

Bicycle	◉ ◉ ◉ ◉
Skates	◉ ◉ ◉
Wagon	◉ ◉
Skateboard	◉

Key: Each ◉ stands for 3 votes.

Test Taking Tips

Get the information you need.
Use the key to find out how many votes each wheel stands for.

1. How many students like skates best?
   - **A** 6 students
   - **B** 9 students
   - **C** 12 students
   - **D** NOT HERE

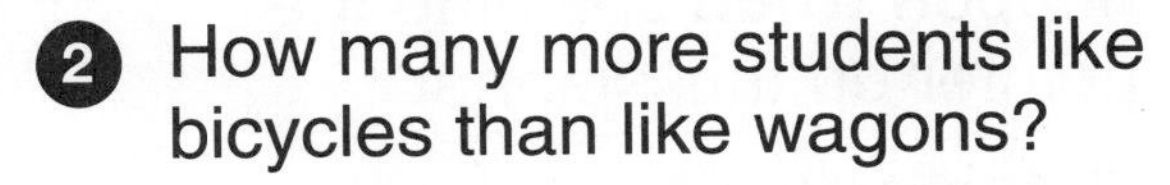

2. How many more students like bicycles than like wagons?
   - **F** 1 more
   - **G** 3 more
   - **H** 6 more
   - **J** 9 more

For questions 3–5, use the table.

CARS IN A CAR SHOW

Kind	Black	White	Red	Blue
Vans	0	2	1	5
Sports Cars	2	1	4	0
Luxury Cars	3	2	1	4

3. How many red sports cars were there?
   A 2 **B** 3 **C** 4 **D** NOT HERE

4. What was the most popular color for the cars there?
   - **F** black
   - **G** blue
   - **H** red
   - **J** white

5. How many vans were white or blue?
   A 6 **B** 7 **C** 8 **D** NOT HERE

For questions 6–8, use the table.

STUDENTS' FAVORITE VACATION

Place	Votes
Beach	卌 卌 |||
Theme Park	卌 ||||
Campground	卌 ||
Home with Friends	卌	

6. How many students answered the survey?
   - **F** 24 students
   - **G** 28 students
   - **H** 30 students
   - **J** 35 students

7. How many more students liked the beach than liked a campground?
   - **A** 2 more
   - **B** 4 more
   - **C** 6 more
   - **D** 8 more

8. What place did the most students like?
   - **F** beach
   - **G** theme park
   - **H** campground
   - **J** home

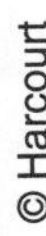

GO ON

Name ______________________________

For questions 9–11, use the bar graph.

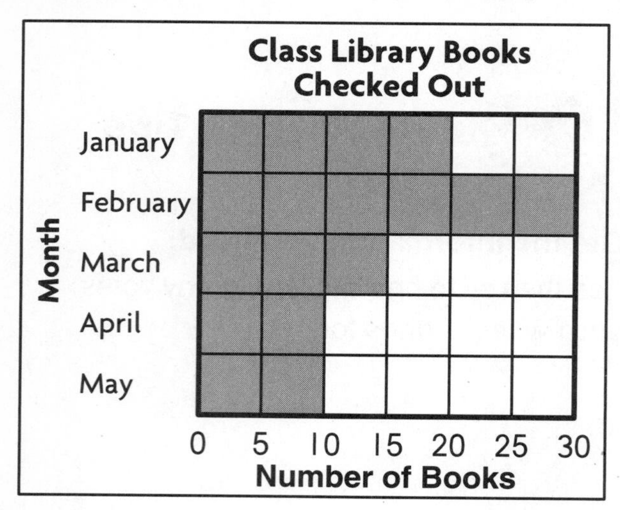

9 How many more books were checked out in February than in April?

A 5 more books
B 10 more books
C 15 more books
D 20 more books

10 In what month were the most books checked out?

F January
G February
H March
J April

11 How many more books were checked out in January than in April?

A 5 more books
B 10 more books
C 15 more books
D 20 more books

12 Which event is possible?

F Someone will read a book.
G You will grow 3 feet in a day.
H Someone will take a boat to the moon.
J The sun will not rise and set.

For questions 13–15, use the spinner.

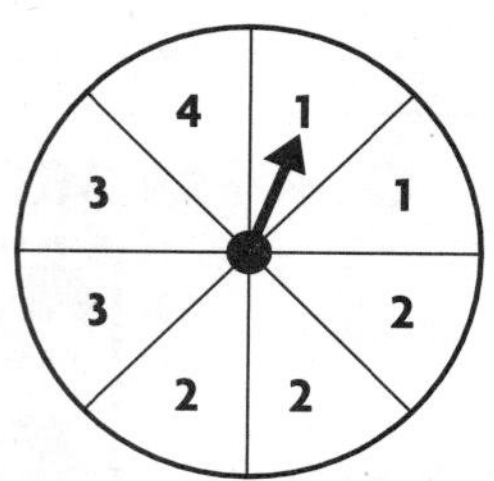

13 Which event is most likely to happen when spinning the pointer on this spinner?

A spinning a 1
B spinning a 2
C spinning a 3
D spinning a 4

14 Which event is least likely to happen when spinning the pointer on this spinner?

F spinning a 1
G spinning a 2
H spinning a 3
J spinning a 4

15 Is the spinner fair?

A yes
B no

16 Carl is playing a game. He has blue marbles, red marbles, and green marbles in a bag. Which color marble is it impossible for Carl to pull out of the bag?

F green
G blue
H red
J yellow

GO ON

Name ______________________________

17 The third graders took a survey to find out what people wanted to eat at the picnic. 16 people wanted sandwiches, 24 wanted fried chicken, 9 wanted hamburgers, and 12 wanted fruit salad.

Complete the table. Use tally marks that show the information.

Third Graders' Favorite Picnic Food	
Sandwiches	
Chicken	
Hamburgers	
Fruit Salad	

On the lines, explain why the table shows the same information as in the paragraph above.

____

____

How can counting by 5's help you check your work?

18 The third graders voted on their favorite pets. The chart shows the results of the vote.

Favorite Pets	Number of Students
Fish	3
Turtles	8
Cats	23
Hamsters	15
Iguanas	11
Rabbits	19
Dogs	38

Jana, a new student, was not there when the others voted. Based on the data in the chart, what do you predict will be Jana's favorite pet? Explain how you decided.

____

____

____

How can making a list help you solve the problem?

GO ON

Name ______________________________

19 The third-graders voted on the kind of chips they wanted at lunch. Here are the results:

Brand of Chips	Number of Votes	Price
Wavy	23	21 cents
Light 'n' Salty	13	36 cents
Crispy	45	27 cents
Goodies	61	23 cents
Toasties	27	18 cents
Ring-a-Ling	48	25 cents

Which three brands should the school order? Circle them.

On the lines below, explain your choice.

____

____

Solve • Short Answer • Explain • Think

Test Taking Tips

Is there more than one right answer?

20 The third graders used tally marks to keep track of the weather for a month. This is what their tally marks looked like:

Weather in May	
Sunny days	卌 卌
Rainy days	卌 III
Windy days	卌 IIII
Cloudy days	IIII

Make a bar graph. Explain how your bar graph shows the same information.

Weather in May

Weather										
Sunny days										
Rainy days										
Windy days										
Cloudy days										
	0 1	2	3	4	5	6	7	8	9	10

Number of Days

What does each box on the bar graph stand for?

Name ______________________________

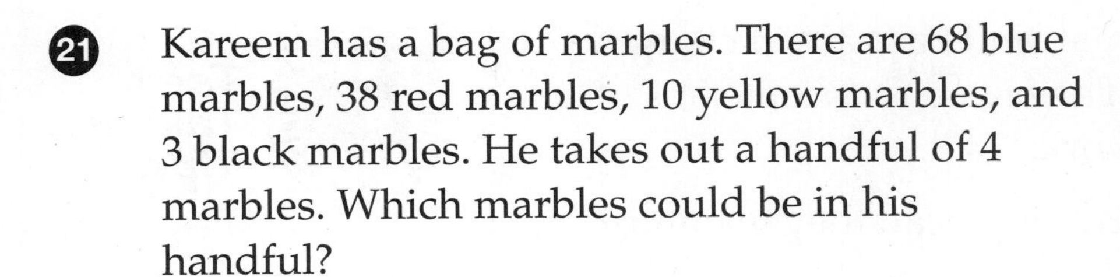

21 Kareem has a bag of marbles. There are 68 blue marbles, 38 red marbles, 10 yellow marbles, and 3 black marbles. He takes out a handful of 4 marbles. Which marbles could be in his handful?

Combination	Possible	Impossible
2 blue, 1 red, 1 black		
4 yellow		
1 green, 3 black		
4 black		

____

____

____

How can making a list of the marbles help solve the problem?

22 Nathan has reading, math, and science homework. He needs to decide which assignment to do first, which to do next, and which to do last.

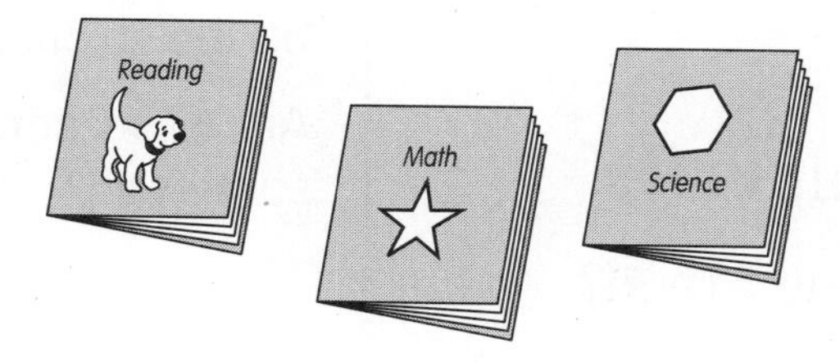

On the lines below, explain what his choices are and how you figured it out.

____

____

____

____

How can making an organized list help you?

GO ON

Name ______________________________

23 Ming-lo, Betty, Jamal, and Sara are in charge of organizing a fair. They need to form teams of two to share the work.

How many different ways can they team up?

____

On the lines below, explain how you solved the problem.

____

____

____

____

Think • Solve • Explain — Test Taking Tips

How can making an organized list help you solve this problem?

24 Brent and Polly were playing a game with this spinner:

Red | Blue | Green | Yellow

1. What are the possible outcomes of spinning on this spinner?
2. Are all possible outcomes equal? Why?
3. What are the chances of spinning green?
4. If Polly and Brent spin 40 times, what do you think their results will be?

Write your answers on the lines below.

____

____

____

____

Think • Solve • Explain — Test Taking Tips

How many sections are on the spinner?

Are all sections the same size?

GO ON

Name ______________________________

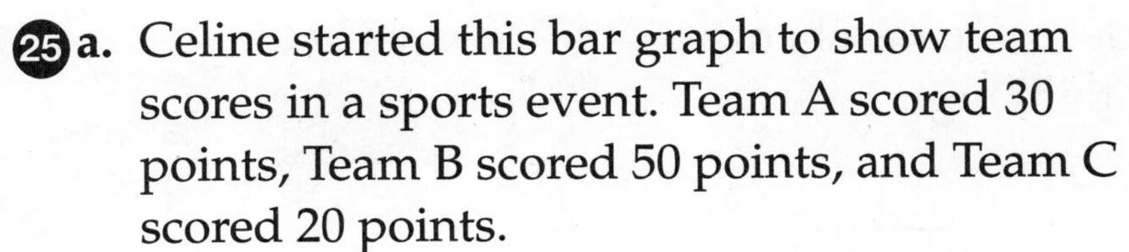

25 a. Celine started this bar graph to show team scores in a sports event. Team A scored 30 points, Team B scored 50 points, and Team C scored 20 points.

Team D scored 5 points more than Team B, Team E scored 20 points more than Team C, and Team F scored 5 points less than Team A.

Figure out the scores for Team D, Team E, and Team F.

Team D Score ______________________

Team E Score ______________________

Team F Score ______________________

Then finish the graph.

Test Taking Tips

What information do you need to complete the graph?

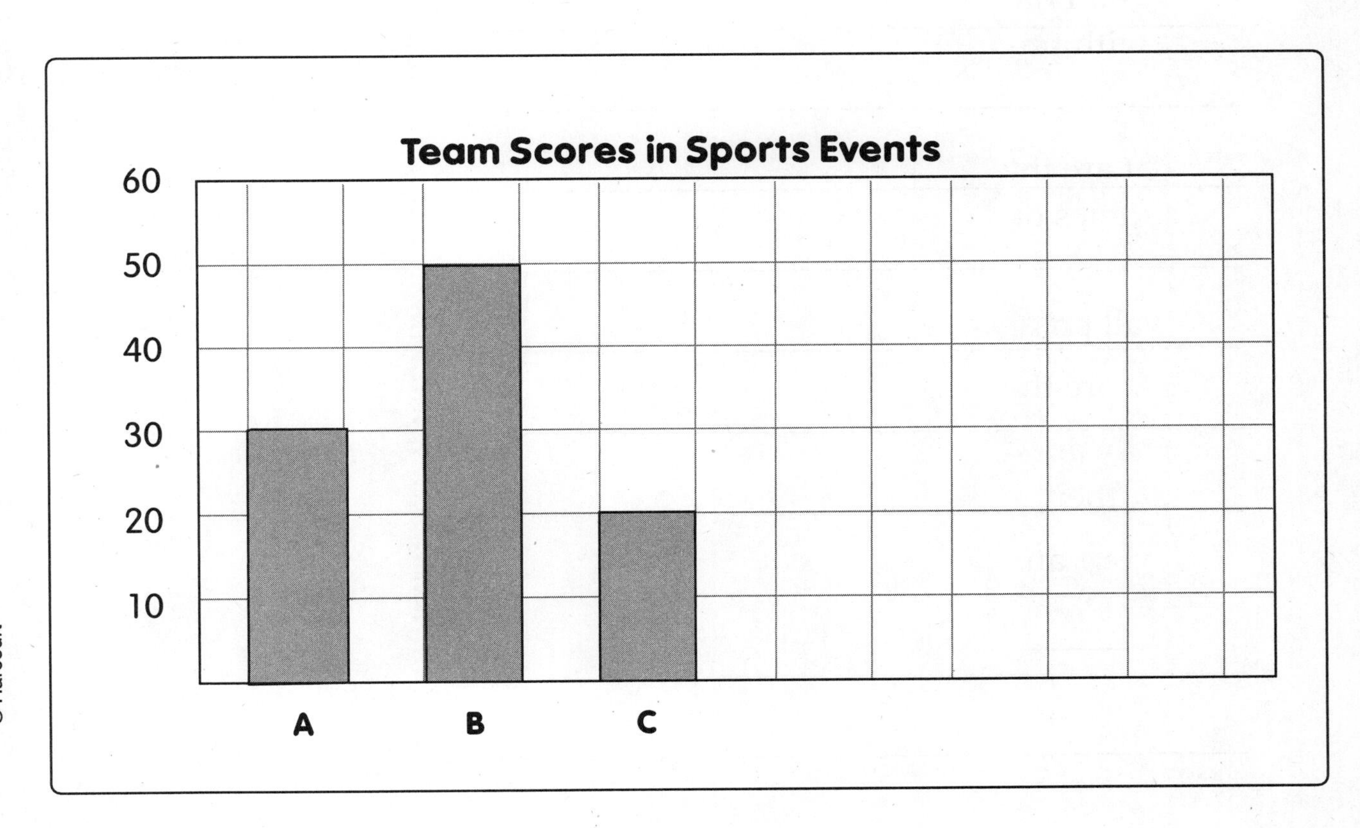

Name ______________________________

25 b. Explain how you decided. Use numbers and words to explain your thinking.

How can you check your answers?

Name ______________________________

26 a. Shaylan's class took a survey to find out how many books students read. They made this chart to show what they found out.

Books Read by Third Graders	
October	𝍸 |
November	𝍸
December	𝍸 𝍸 |
January	𝍸 |||

Make a pictograph that shows the information about books read by third graders.

In your pictograph, use this symbol to stand for 2 books:

□□

Be sure that your pictograph has a title and a key.

What symbol can you use to stand for less than 2 books?

Name ______________________________

26 b. Draw your pictograph here.

Books Read by Third Graders	

Write two sentences that compare information about books read by third graders.

____

____

____

____

____

____

____

____

Solve • Explain • Think
Long Answer

Test Taking Tips

How can you check your answer?

STOP

Name ______________________________

Choose the letter of the correct answer.

1 Is the figure formed by only straight lines, only curved lines, or both straight and curved lines?

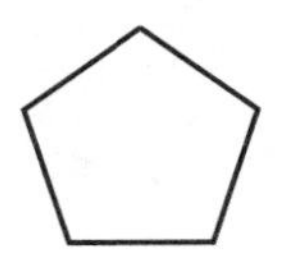

A only straight lines
B only curved lines
C both curved and straight lines

Test Taking Tips

Look for important words.
What key words could you look for to give you clues?

---

2 Identify the solid figure that is like the object shown.

F sphere
G cone
H cube
J cylinder

3 Which solid figure is like a book?

A cylinder
B sphere
C cone
D NOT HERE

4 Which solid figure has the face shown?

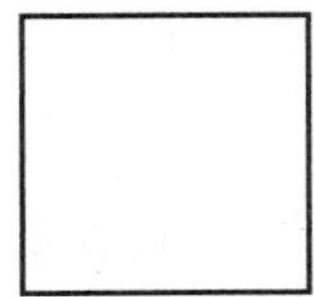

F cylinder
G cube
H cone
J NOT HERE

5 Which figure answers the riddle?
I am a solid figure with no edges and no faces. What am I?

A cube
B square prism
C sphere
D NOT HERE

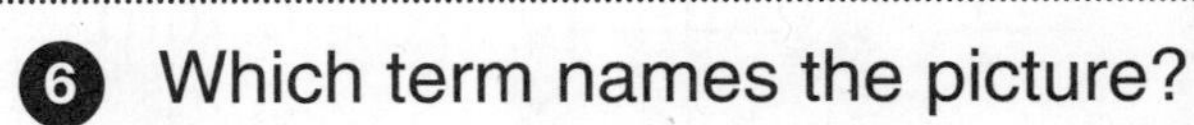

6 Which term names the picture?

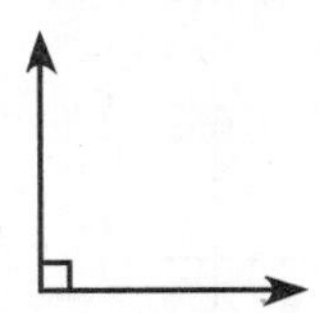

F line
G line segment
H triangle
J right angle

7 How many line segments are in a square?

A 2 **B** 3 **C** 4 **D** 5

For questions 8–9, use the grid showing the design of a kitchen. What object is located at each ordered pair?

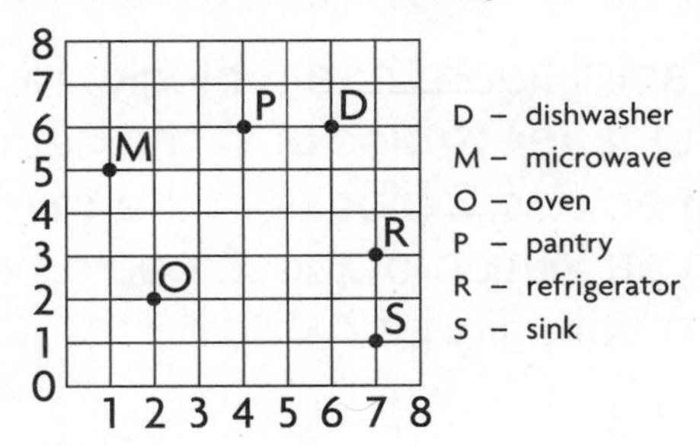

8 (1,5)

F oven
G pantry
H microwave
J dishwasher

9 (6,6)

A dishwasher
B pantry
C refrigerator
D sink

GO ON

Name ______________________________

For questions 10–11, decide which motion was used to move the plane figure.

10 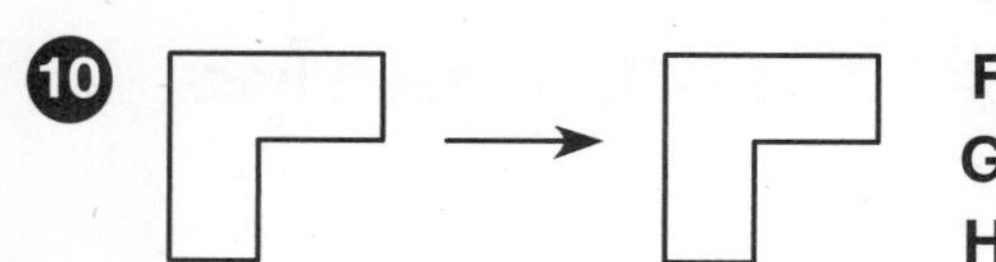

F slide
G flip
H turn

11 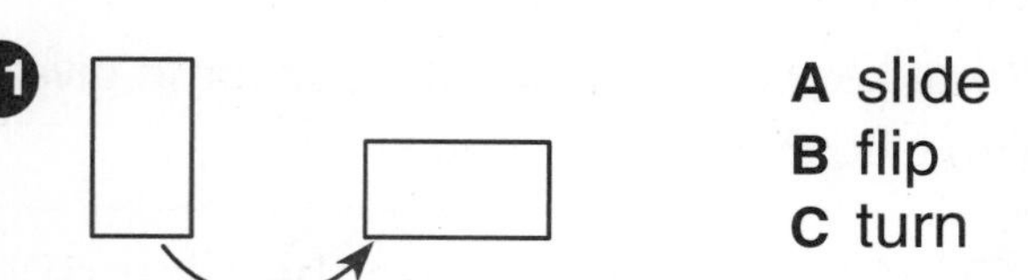

A slide
B flip
C turn

12 What is the next shape in the pattern?

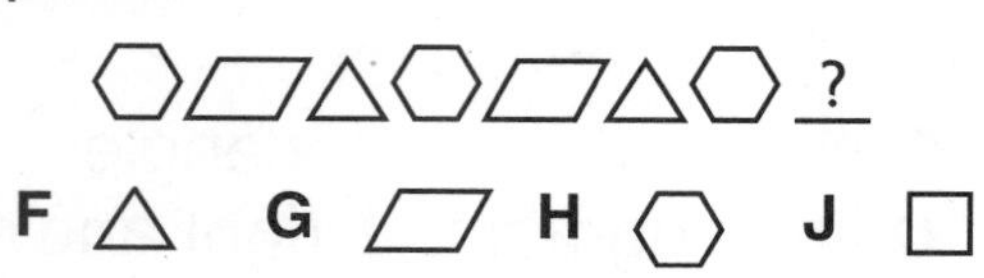

F **G** **H** **J**

13 Which figure is congruent to the figure shown?

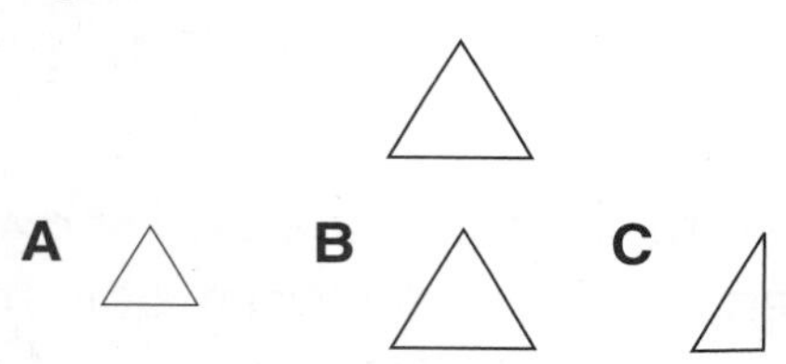

A **B** **C**

14 A solid figure has 5 blocks in the first layer, 4 blocks in the second layer, and 3 blocks in the third layer. Which of these could be the solid figure?

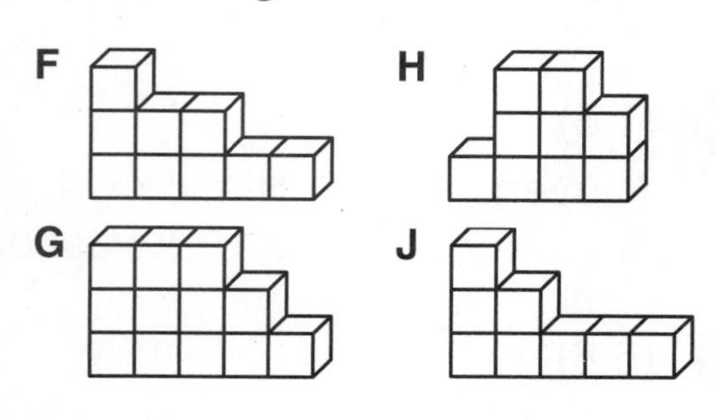

F **H** **G** **J**

15 How many lines of symmetry does the figure have?

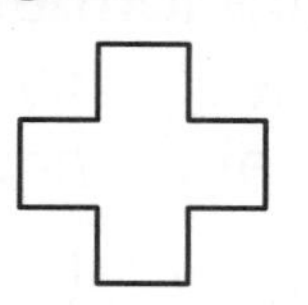

A 0 **B** 1 **C** 4 **D** 5

16 Suki drew this figure. She matched it to make the other half. How does the completed figure look?

F 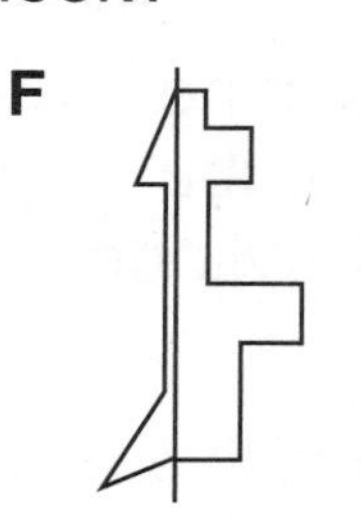

H 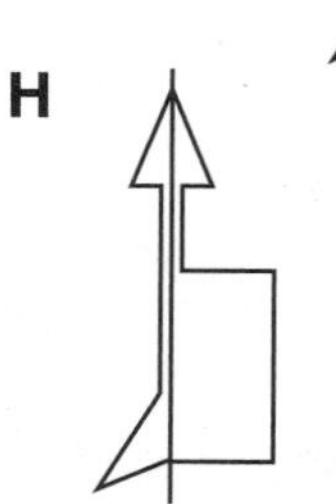

G 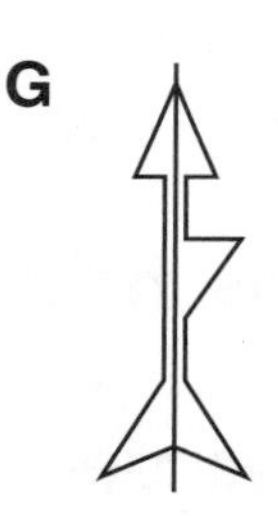

J 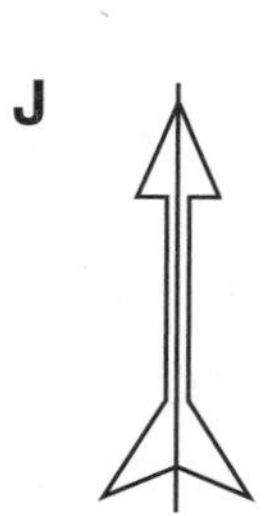

17 Which letter does not have a line of symmetry?

A A
B O
C R
D W

18 Are the solid figures congruent?

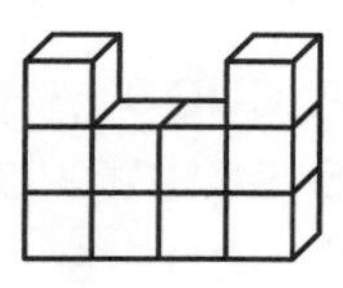 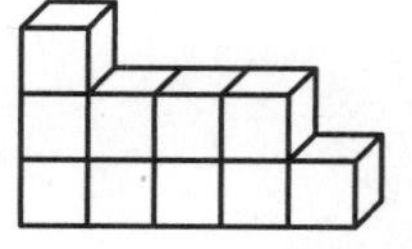

F yes **G** no

GO ON

Name ________________________________

Find the name of a geometric solid in the Word Box to describe each object. Write the name on the line. Use each word once.

Word Box

cylinder cone
sphere rectangular prism

A. a cereal box

____

B. a soccer ball

____

C. a Native American tepee ______________

D. a roll of pennies ______________

In the space below, draw and label a picture of each shape.

How can picturing the objects in your mind help you solve the problem?

Describe each shape. Complete the table.

Shape	Number of Sides	Number of Corners
Triangle		
Rectangle		
Square		

On the lines below, explain how two triangles or two rectangles or two squares can be different from each other.

____

____

____

How can drawing a picture help you solve this problem?

Name ______________________________

21 Connor made a design with a triangular pattern of stars. Each row of the triangle had one less star than the one below it. There was one star in the top row. There were 4 rows in all. How many stars were in the triangle?

On the lines below, explain how you solved the problem.

____

____

____

____

Think · Solve · Explain — Short Answer

Test Taking Tips

How can drawing a picture help you solve the problem?

22 Look at the pattern. Draw the next two shapes that would continue the pattern.

○ △ △ △ ○ △ △ △

On the lines below, explain how you decided what the next two shapes should be. In your explanation, use the names of the shapes.

____

____

____

____

Think · Solve · Explain — Short Answer

Test Taking Tips

How can you find the pattern that repeats?

Name ____________________________________

Draw three squares on the dot paper below. Make two of them congruent. Then circle the two squares that are congruent.

On the lines below, explain why the two squares you circled are congruent. Then explain why the other one is not congruent.

____

____

____

Test Taking Tips

What math word do you need to know? How can the dot paper help you draw figures that are the same size and shape?

Look at each pair of figures. Are they congruent?

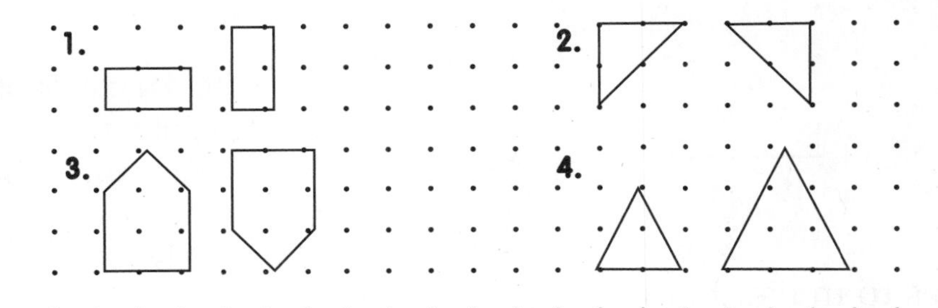

Write your answers on the lines below. Explain why each pair is congruent or is not congruent.

____

____

____

Test Taking Tips

What is true of congruent figures?

Name ______________________________

The parking deck has a place to park motorcycles. Every space is filled. Ray's motorcycle is in the middle. There are 5 motorcycles to the right of Ray's.

How many motorcycles are there in all?

Draw a picture that shows how the motorcycles are parked. Use an X to represent each motorcycle. Label Ray's motorcycle.

Then explain the steps you followed to figure out the answer.

____

____

In your picture, what do you need to show first?

What important detail do you know that will help you solve the problem?

Opal drew the first half of these figures.

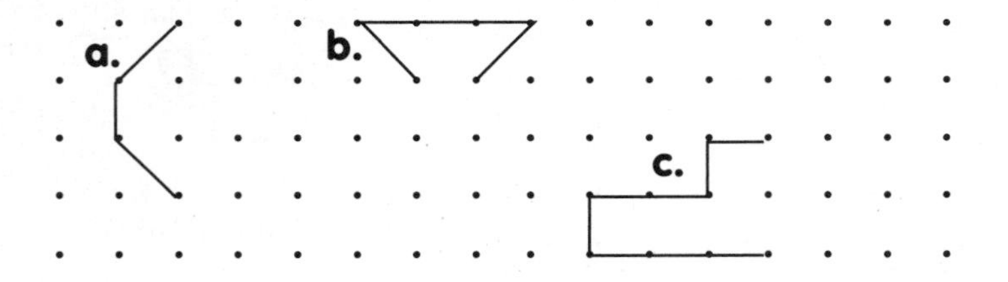

Complete the figures to make them symmetrical. Then draw all the lines of symmetry that you can find in each figure.

On the lines below, explain why each of your finished figures is symmetrical.

____

____

If you held a mirror up to each unfinished edge, what would you see?

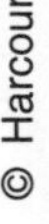

GO ON

Name ______________________________

27 a. Which shape could you make with each set of sticks? Write the letter:
A. rectangle B. triangle C. square

1
4 in.
1 in.
3 in.
____

2
2 in. 2 in.
2 in. 2 in.
____

3
4 in.
4 in.
4 in.
____

4
3 in.
3 in.
1 in.
____

5
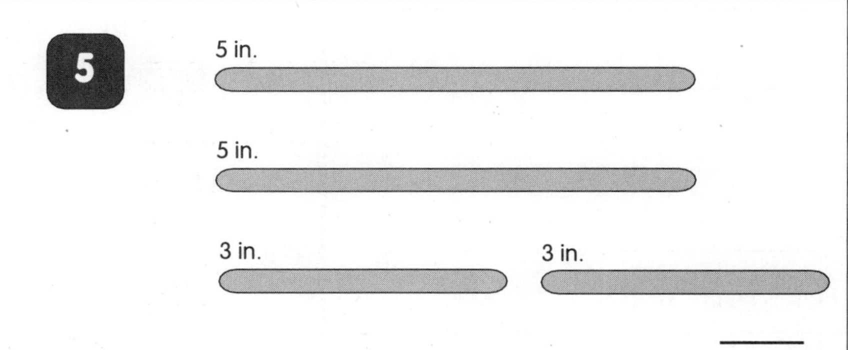

____

Test Taking Tips

Think • Solve • Explain — Long Answer

What do you know about rectangles, squares, and triangles?

GO ON

Name ____

27 b. Draw each shape on this page. Name the shape.

Explain how you know what shape could be made with each set of sticks.

____

____

____

____

____

How can making a model with sticks help you solve the problem?

GO ON

Name ______________________________

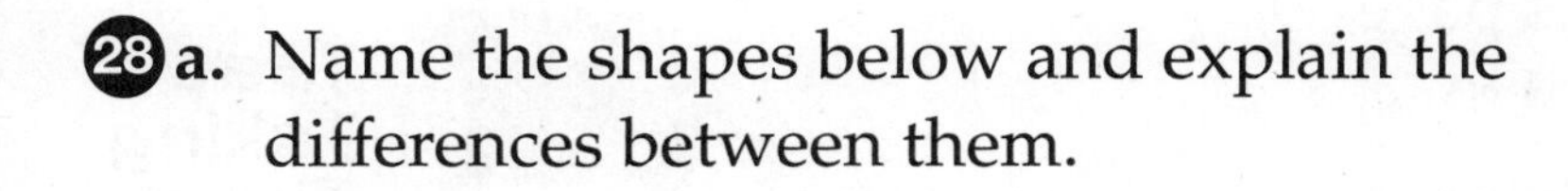

28 a. Name the shapes below and explain the differences between them.

A

B

In your explanation, remember to include the number of sides, the size of the corners, and the length of the sides.

Test Taking Tips

What is alike about the shapes? What is different about them?

Name ______________________________

28 b. Write your answer on this page.

How can you check that your explanation is clear and complete?

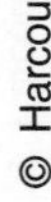

Name ______________________________

Choose the letter of the correct answer.

For questions 1–2, use the picture.

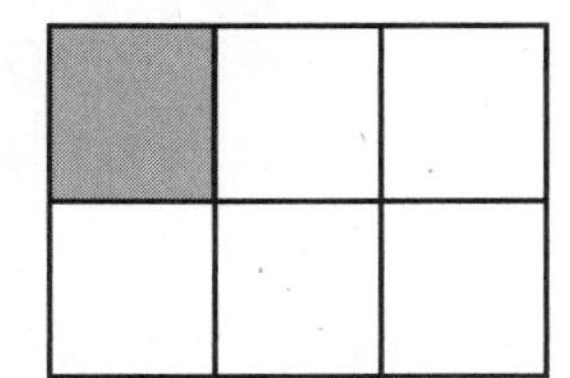

Test Taking Tips

Understand the problem.

Count the total number of parts in the fraction model.

1 How many parts make up the whole?

A 4 **B** 6 **C** 8 **D** 10

2 How many parts are shaded?

F 1 part
G 2 parts
H 5 parts
J 6 parts

For questions 3–4, find the numbers or words that name the part of the group that is shaded.

3

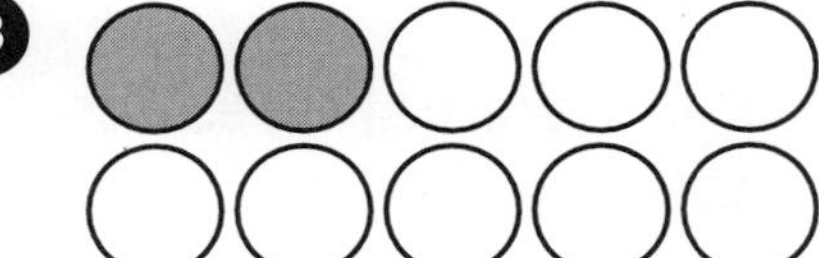

A $\frac{1}{10}$ **B** $\frac{1}{8}$ **C** $\frac{1}{6}$ **D** $\frac{1}{5}$

4

F one eighth
G one sixth
H one fourth
J one half

5 What is the fraction in numbers?

5 out of 8

A $\frac{1}{10}$ **B** $\frac{5}{8}$ **C** $\frac{5}{6}$ **D** $\frac{3}{4}$

6 Choose $<$, $>$, or $=$ to compare.

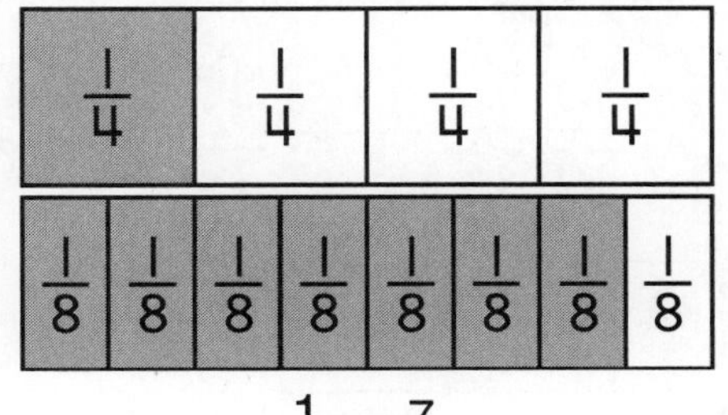

$\frac{1}{4} \bullet \frac{7}{8}$

F $<$ **G** $>$ **H** $=$

7 Kelly used $\frac{4}{8}$ foot of ribbon and Danielle used $\frac{3}{4}$ foot of ribbon to wrap a gift. Who used more ribbon?

A Kelly **B** Danielle

8 Mr. Dobins has 8 hot dogs to grill. He has grilled 4 hot dogs so far. What part of the hot dogs are grilled?

F $\frac{3}{8}$ **G** $\frac{4}{8}$ **H** $\frac{5}{8}$ **J** $\frac{6}{8}$

9 Jim poured milk in 4 of 12 cups. What part of the cups had milk?

A $\frac{2}{12}$ **B** $\frac{4}{12}$ **C** $\frac{6}{12}$ **D** $\frac{8}{12}$

GO ON

Name ______________________________

10 Which decimal names the shaded part?

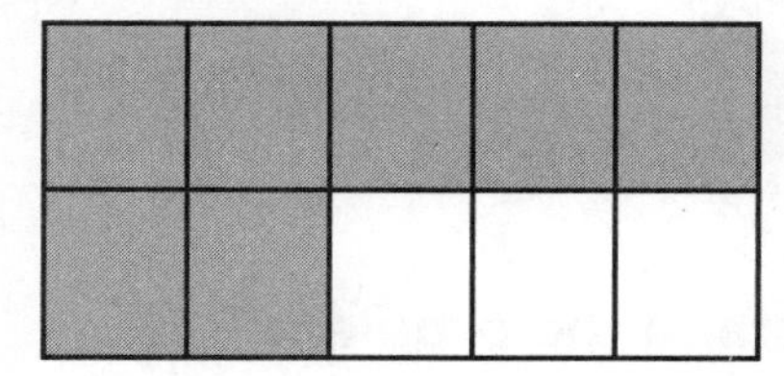

F one tenth
G seven tenths
H ten sevenths
J NOT HERE

11 Compare the parts of the group that are shaded. Choose <, >, or =.

$\frac{5}{8} \bullet \frac{7}{8}$

A < **B** > **C** =

12 Choose <, >, or =.

$\frac{2}{3} \bullet \frac{2}{3}$

F < **G** > **H** =

13 What is $\frac{6}{10}$ written as a decimal?

A 0.10
B 0.6
C 6.0
D 10.6

14 What is 0.7 written as a fraction?

F $\frac{1}{10}$
G. $\frac{7}{10}$
H $\frac{10}{10}$
J. $\frac{10}{7}$

15 What is thirty-eight hundredths written as a decimal?

A 0.38
B 3.80
C 38.0
D NOT HERE

16 What are the words for the mixed decimal 10.62?

F ten and sixty-two hundredths
G ten and six halves
H one hundred sixty-two
J NOT HERE

17 Anna ate 0.2 of her cookies and saved 0.8 of them. Did she eat more or save more?

A She ate more.
B She saved more.

18 What mixed decimal does the model show?

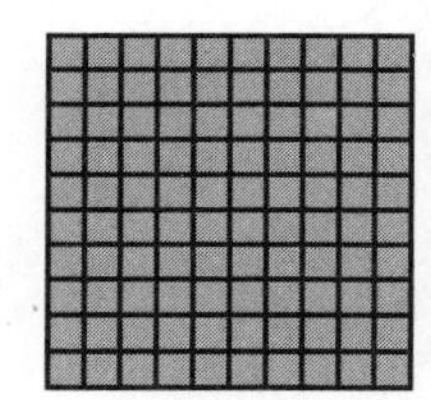

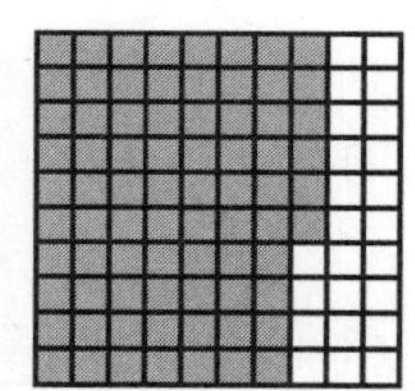

F 0.76
G 1.76
H 2.76
J 7.90

19 Wade jogged 1.8 miles on Tuesday and 1.6 miles on Thursday. On which day did he jog farther?

A Tuesday
B Thursday

Name ______________________________

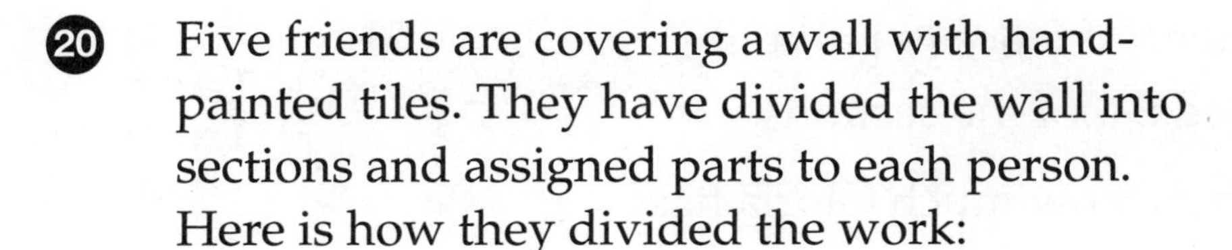

20 Five friends are covering a wall with hand-painted tiles. They have divided the wall into sections and assigned parts to each person. Here is how they divided the work:

Nate	Joe
Joe	Sasha
Joe	Sue
John	Sasha

On what fractional part of the wall is Joe working?

On the lines below, explain how you decided.

____

____

____

How many parts are there in all?

21 Cassie sorted her beads in a special box. This is how she arranged them:

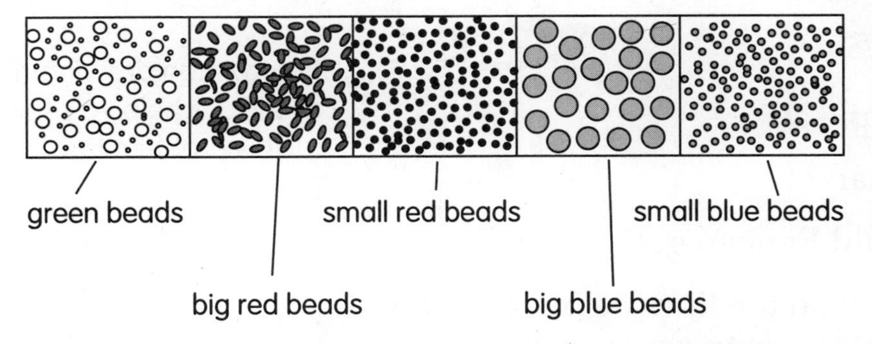

What fraction of the box holds blue beads? Use numbers and words. On the lines below, explain how you decided.

____

____

How many equal sections does the bead case have?

How many kinds of blue beads does Cassie have?

Name ______________________________

 Gloria had five sections of fence to paint. By noon she had finished $3\frac{1}{3}$ of the sections. Draw a picture to show how much Gloria has painted.

On the lines below, explain how you decided.

____

____

____

Think • Solve • Explain
Short Answer
Test Taking Tips

How many painted sections are completely painted? How many are partly painted?

23 Reuben is having a birthday party. He says guests will be sharing mini-pizzas. Each guest will get $1\frac{1}{4}$ pizzas.

If there are 4 people at the party, how many mini-pizzas should Reuben get?

Write your answer on the lines below. Explain how you solved the problem.

____

____

____

How can drawing a picture help you solve this problem?

Name ______________________________

 Quentin has finished $\frac{2}{5}$ of his math homework.

Fiona has finished $\frac{2}{3}$ of the same assignment.

Who has finished more homework?

On the lines below, explain your reasoning.

____

____

____

____

____

____

Solve • Explain • Think — Short Answer

Test Taking Tips

How can drawing a picture for each fraction help you decide?

 Jan paints 7 paper plates.

She paints 4 plates blue.

She paints the other plates red.

What fraction of the plates are red?

Use pictures and words to explain your thinking.

Test Taking Tips

How can drawing a picture help you see the whole group and its parts?

GO ON

Name ______________________________

Jake folded newspapers for his uncle. The first day, he folded 100 newspapers and was paid $5.00. The second day, he folded 150 newspapers and was paid $7.50. The third day, he folded 200 newspapers and was paid $10.00. On the fourth day, he folded 250 newspapers.

What do you think Jake was paid on the fourth day?

On the lines below, explain how you decided.

____

____

____

What is the pattern?

Each area of the parking lot has 10 spaces. The shaded parts of the model show how many parking spaces are taken.

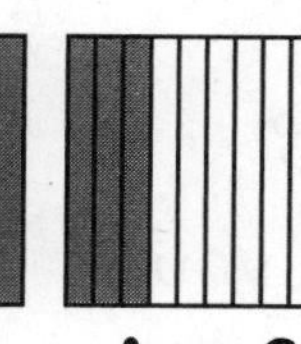

Area A   Area B   Area C

Write a mixed decimal that shows how many areas of the parking lot are filled.

On the lines below, explain how you figured it out.

____

____

What is a mixed decimal?

Name ______________________________

a. George ate half of a pizza. Corinne ate half of a different pizza. George said that he ate more than Corinne. Corinne said they both ate the same amount because they each ate half of a pizza.

Could George be right? ____________________

Test Taking Tips

When is half of one thing more than half of another?

28 b. Use words and pictures to make your answer clear.

____

____

____

____

____

____

____

____

____

____

____

____

____

____

____

Solve • Think • Explain — Long Answer

Test Taking Tips

Be sure that your explanation is clear and complete.

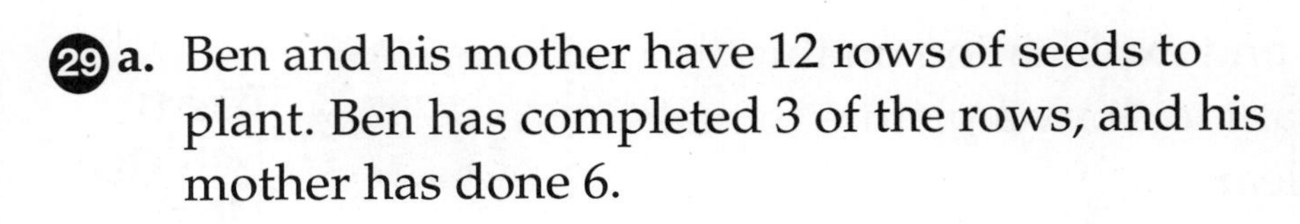

Name ______________________________

29 **a.** Ben and his mother have 12 rows of seeds to plant. Ben has completed 3 of the rows, and his mother has done 6.

Together, what fractional part of the total have they completed?

____

What do you need to find out first?

GO ON

Name ______________________________

29 b. Write your answer and explanation on this page. You might want to draw a picture to make your explanation clear.

____

____

____

____

____

____

____

____

____

____

____

____

How can you check your answer?

STOP

Name ______________________________

Choose the letter of the correct answer.
You will need a ruler for some of the questions.

For questions 1–2, choose the best unit of measure.

1. A backpack is about 15 __?__ wide.
   A inches
   B feet
   C yards
   D miles

Test Taking Tips

Eliminate choices.
Which units of measurement would not be good choices?

2. Most walls are about 8 __?__ high.
   F inches
   G feet
   H yards
   J miles

For questions 3–4, measure the length to the nearest half inch.

3. (line segment)
   A 1 in.
   B $1\frac{1}{2}$ in.
   C 2 in.
   D $2\frac{1}{2}$ in.

4. 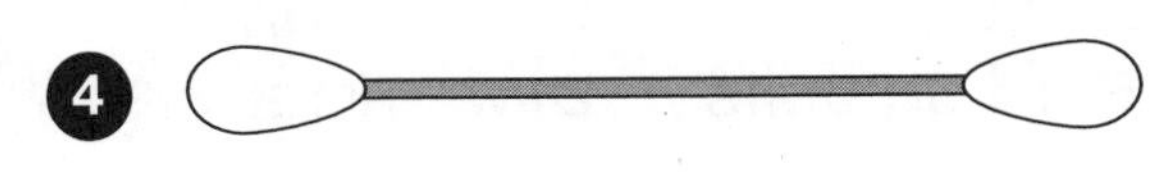
   F $1\frac{1}{2}$ in.
   G 2 in.
   H $2\frac{1}{2}$ in.
   J 3 in.

5. Brad measured the distance from the ice cream shop to the ski shop across the street. Which of these was the distance between the two shops?
   A 10 dm   B 10 cm   C 10 m

6. Which is the better estimate of a doughnut?
   F 1 ounce   G 1 pound

7. What is the best estimate?

   A 18 cups
   B 18 pints
   C 18 quarts
   D 18 gallons

8. Which unit of measure should be used to measure the width of a dime?
   F centimeter
   G decimeter
   H meter

9. Which is the better estimate of capacity?

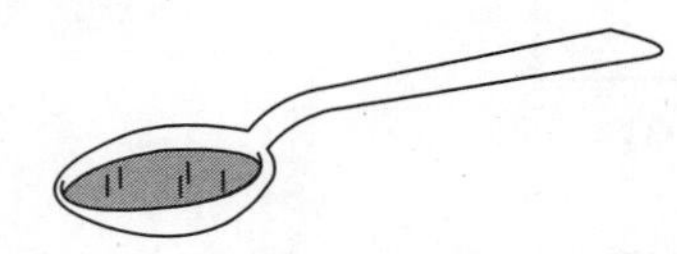

   A 5 mL   B 5 L

GO ON

Name ____________________

10 Which is the better unit to measure the weight of a bowling ball?

F g
G kg

For questions 11–12, use a centimeter ruler to measure the length.

11 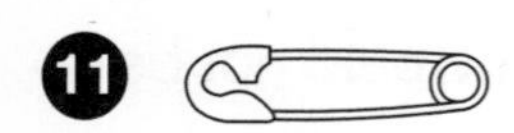

A 1 cm　　**C** 3 cm
B 2 cm　　**D** 4 cm

12 

F 1 cm　　**H** 3 cm
G 2 cm　　**J** 4 cm

For questions 13–14, find the perimeter of the figure.

13 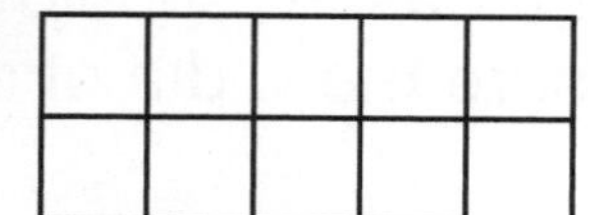

A 8 units　　**C** 12 units
B 10 units　　**D** 14 units

14 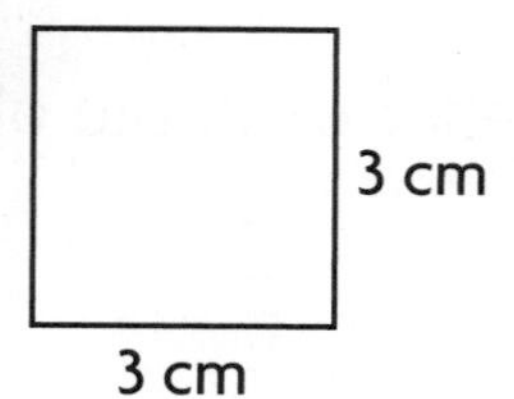

F 4 cm　　**H** 10 cm
G 8 cm　　**J** 12 cm

15 What is the perimeter of the figure?

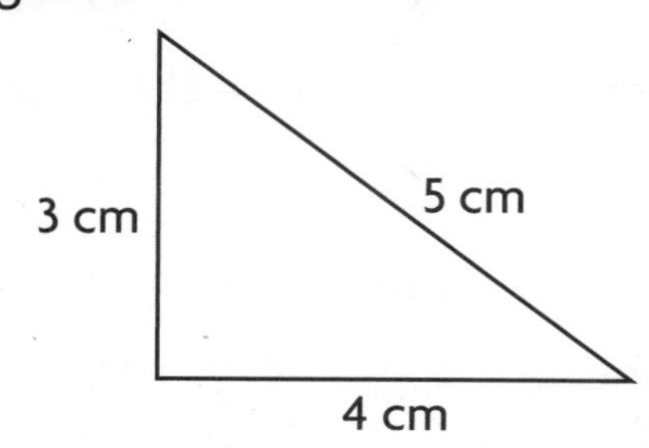

A 3 cm　　**C** 9 cm
B 6 cm　　**D** 12 cm

16 Tom is replacing a wood frame around a window that is 7 feet high and 3 feet wide. How many feet of wood does he need?

F 10 ft　　**H** 20 ft
G 15 ft　　**J** 21 ft

17 What is the area of the figure in square units?

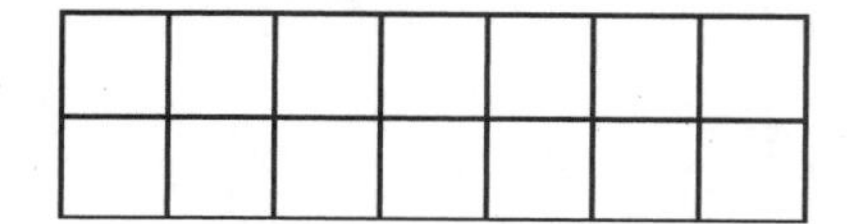

A 9 sq units　　**C** 18 sq units
B 14 sq units　　**D** NOT HERE

18 A rectangle has an area of 28 square feet and a side 7 feet long. What is its perimeter?

F 20 ft　　**H** 22 ft
G 24 ft　　**J** 26 ft

19 A rectangular pizza is 8 units long and 4 units wide. How many square units can be cut from it?

A 24 sq units　　**C** 32 sq units
B 28 sq units　　**D** 36 sq units

GO ON

Name ______________________________

Patrick drove 34 miles to get to the beach. He stayed there for 3 hours and then drove home. Natalie drove 68 miles to get to the beach. She stayed there for 2 hours before driving home. How many more miles did Natalie drive than Patrick?

On the lines below, explain how you solved the problem.

____

____

____

____

What information do you need to solve the problem?

Mei wants to make a poster showing five sharks. She wants to put them in order, with the longest one on top and the shortest one on bottom.

	Length of Sharks	
	Horn Shark	4 ft
	Thresher Shark	15 ft
	Great White Shark	18 ft
	Basking Shark	25 ft
	Mako Shark	9 ft

Number the sharks from longest to shortest.

On the lines below, explain how you know.

____

____

____

What are you trying to find out? Restate the problem in your own words.

GO ON

Name ________________________________

22 There will be eight children at Megan's party. Megan's mother will serve juice. She predicts that each child will drink about two cups of juice.

How many QUARTS of juice will she need to make?

On the lines below, show how you solved the problem. Use numbers and words.

____

____

____

Solve • Think • Explain — Short Answer

Test Taking Tips

How many cups are in a quart?

23 On Tuesday it was 85° at noon.

On Wednesday, the temperature was 10 degrees cooler. Show Wednesday's temperature on the blank thermometer.

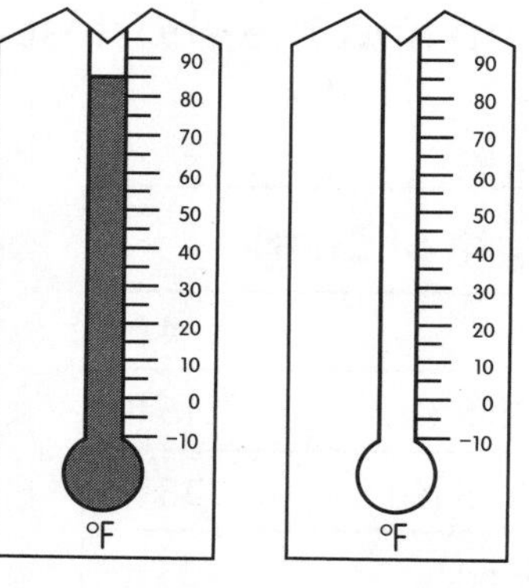

On the lines below, explain the steps you followed to figure out what the temperature was on Wednesday.

____

____

____

When it's cooler, is the temperature higher or lower?

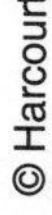

GO ON

Name ____________________________________

Vinh made a bar graph showing the high temperatures for six days.

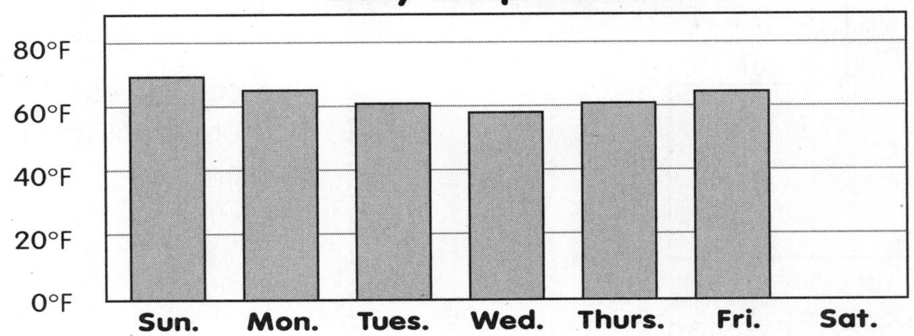

What was the highest temperature of the week? What was the lowest?

What do you predict the temperature might be on Saturday?

On the lines below, explain how you decided.

____

____

Rhonda's room is 10 feet by 12 feet. She wants to put a strip of wallpaper trim around the walls near the ceiling. How many feet of wallpaper trim will she need?

On the lines below, explain how you figured out the answer.

____

____

____

____

____

Solve · Think · Explain · Short Answer

Test Taking Tips

How can drawing a picture help you solve the problem?

GO ON

Name ____

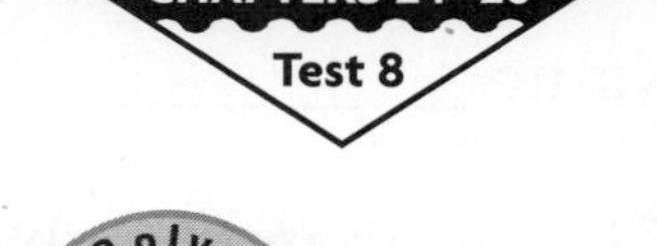

26 The picture shows the playground at Willow Park. What is the perimeter of the playground?

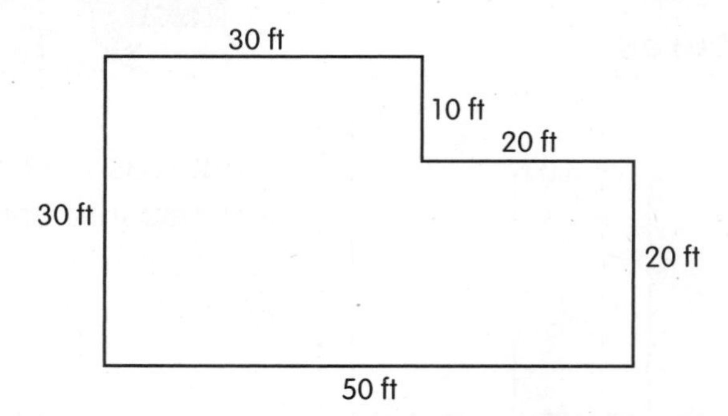

Remember, perimeter is the distance around a figure.

Write a number sentence you can use to solve the problem.

____

How can you use the diagram to solve the problem?

27 Randi leaves home at 3:30 P.M. and walks to the library. She spends half an hour there. Then she walks to the music store for her one-hour guitar lesson. After that, she walks home. It takes Randi 15 minutes to walk half a mile, or 30 minutes to walk a mile.

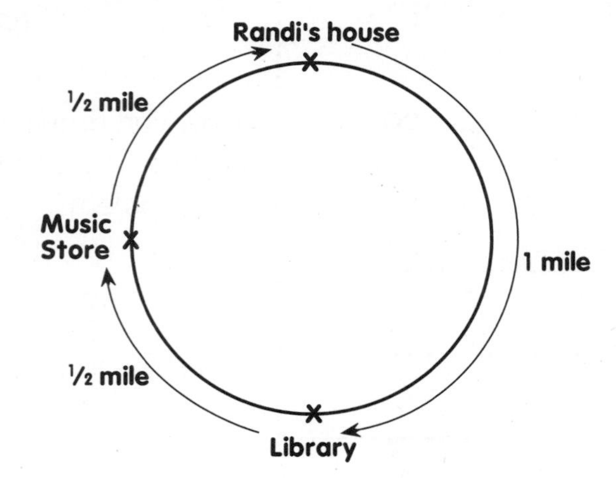

What time does Randi get home?

Explain how you know. ___________________________

What time does Randi get to the library? When does she leave?

Name ______________________________

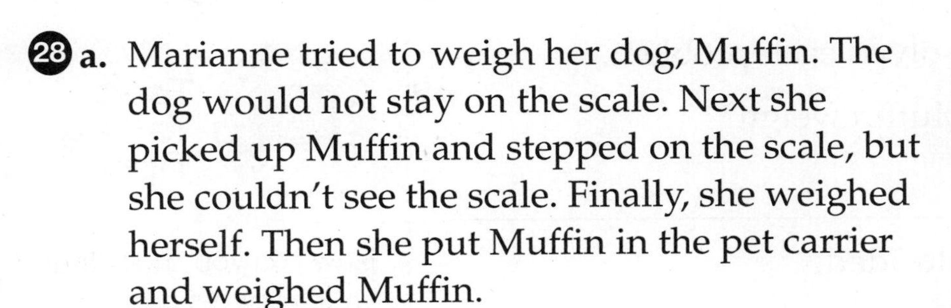

28 **a.** Marianne tried to weigh her dog, Muffin. The dog would not stay on the scale. Next she picked up Muffin and stepped on the scale, but she couldn't see the scale. Finally, she weighed herself. Then she put Muffin in the pet carrier and weighed Muffin.

The chart shows all the information she has gathered.

On the Scale	Pounds
Marianne	65
Muffin the dog	
pet carrier	5
Marianne and Muffin	
Muffin in the pet carrier	30

Use the chart. How much does Muffin weigh?

How much do Muffin and Marianne weigh together?

What facts on the chart will help you solve the problem?

GO ON

Name ______________________

28 b. Explain how you solved each problem.

How much does Muffin weigh?

____

Explain how you decided.

____

____

____

____

____

How much do Marianne and Muffin weigh together?

____

Explain how you decided.

____

____

____

____

____

____

____

____

How can you check that your answers are correct?

How can you check that your explanations are clear?

Name ______________________________

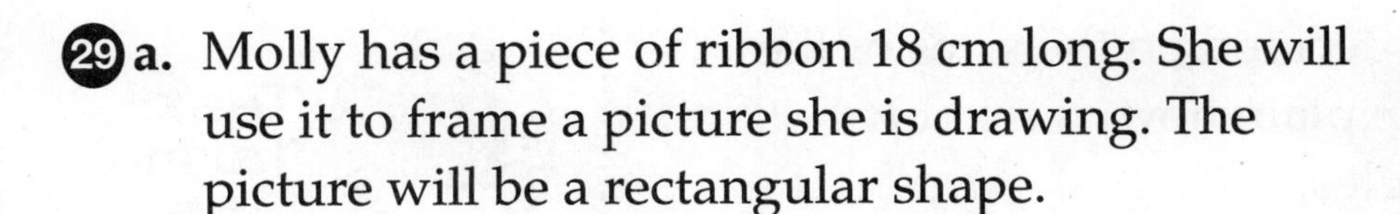

29 **a.** Molly has a piece of ribbon 18 cm long. She will use it to frame a picture she is drawing. The picture will be a rectangular shape.

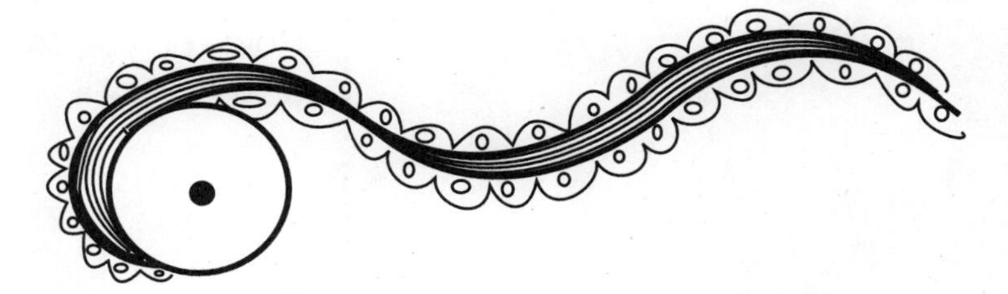

Show two different-sized rectangular shapes that Molly could use for her drawing. Label each side with the number of centimeters.

Test Taking Tips

How can using the strategy Guess and Check help you solve the problem?

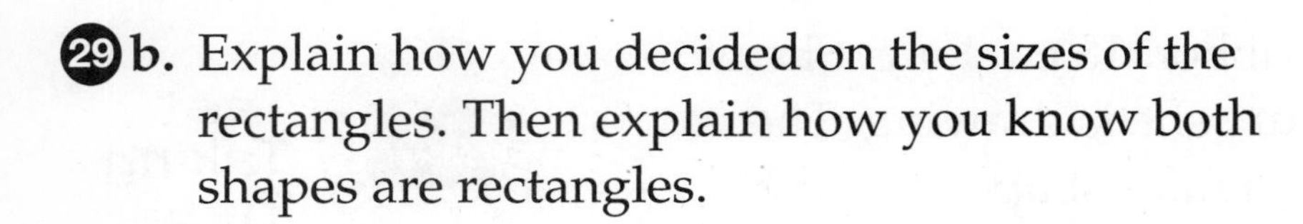

29 b. Explain how you decided on the sizes of the rectangles. Then explain how you know both shapes are rectangles.

Solve · Explain · Think
Long Answer

Test Taking Tips

How can you check your answer?

STOP

Name ______________________________

Choose the letter of the correct answer.

For questions 1–4, find the product. Use base-ten blocks.

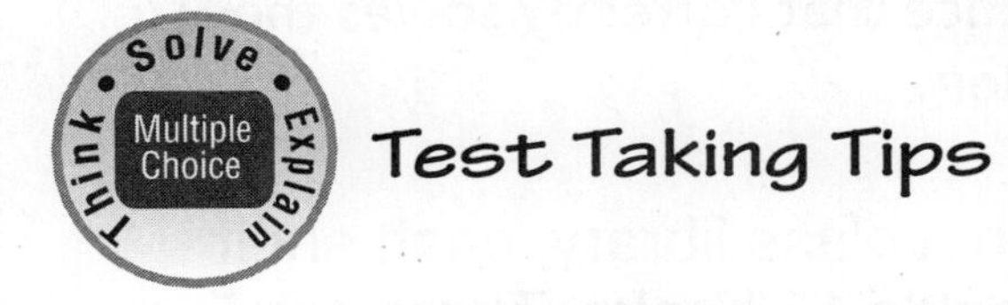

Understand the problem.
How many rows of base-ten blocks would you need? How many blocks would be in each row?

1 32×6

A 182 C 192 E NOT HERE
B 188 D 198

2 $9 \times 71 = \underline{?}$

F 161 H 638 K NOT HERE
G 169 J 639

3 $9 \times 14 = \underline{?}$

A 96 C 136 E NOT HERE
B 126 D 138

4 44×5

F 200 H 225 E NOT HERE
G 220 J 245

5 Mrs. Lane's car goes 26 miles for each gallon of gasoline. Today she used 7 gallons. How many miles did she travel?

A 182 miles
B 242 miles
C 282 miles
D 1,413 miles
E NOT HERE

6 $47 \div 4 = \underline{?}$

F 4 r7 H 11 r4 K NOT HERE
G 11 r3 J 12

7 $49 \div 4 = \underline{?}$

A 12 r1 C 13 r1 E NOT HERE
B 12 r2 D 13 r3

8 $37 \div 2 = \underline{?}$

F 19 r2 H 18 r2 K NOT HERE
G 19 r1 J 18 r1

9 $56 \div 6 = \underline{?}$

A 9 C 9 r2 E NOT HERE
B 10 r6 D 9 r6

10 Mrs. Johnson had 50 balloons. She gave the same number to each of 7 children at the party. How many balloons did each child get? How many balloons were left over?

F 7 balloons each; 0 left over
G 7 balloons each; 1 left over
H 8 balloons each; 1 left over
J 8 balloons each; 3 left over
K NOT HERE

Name ___________________________________

For questions 11–12, choose the number sentence that correctly solves the problem.

11 In a class library, each shelf holds 15 books. There are 5 shelves. How many books are in the class library?

A $15 \div 5 = 3$ **C** $15 + 5 = 20$
B $15 - 5 = 10$ **D** $5 \times 15 = 75$

12 Each student in a class reads 4 books. There are 23 students in the class. How many books do the students read?

F $4 \times 23 = 92$
G $23 + 4 = 27$
H $23 - 4 = 19$
J $23 \div 4 = 5$ r3

For questions 13–14, use the array. Add the two products to find the answer that completes the multiplication sentence.

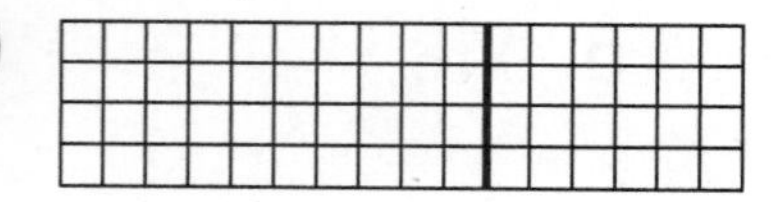

$3 \times 16 =$ ___?___

A 38 **C** 49
B 48 **D** NOT HERE

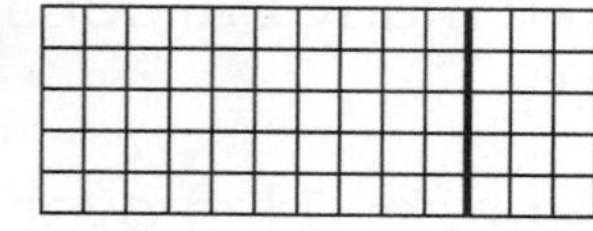

$5 \times 13 =$ ___?___

F 30 **H** 55
G 50 **J** 65

15 A party table has 6 trays on it. Each tray holds 15 cookies. How many cookies are on the table?

A 60 cookies **C** 95 cookies
B 90 cookies **D.** 100 cookies

16 Ben's garden has 12 plants across each row. How many plants are there in 6 rows?

F 18 plants
G 62 plants
H 72 plants
J 78 plants

17 Lucy is making a quilt. The quilt will have 7 rows, with 12 squares in each row. What will the area of the quilt be?

A 84 sq units **C** 94 sq units
B 89 sq units **D** 104 sq units

For questions 18–19, choose whether to multiply or divide. Solve the problem.

18 A paint box has 42 tubes of paint. If 7 children share the paints, how many tubes will each child have?

F divide; 6 tubes
G divide; 9 tubes
H multiply; 294 tubes
J multiply; 296 tubes

19 Tara bought 3 skirts. Each skirt cost \$19. How much did Tara spend on the skirts?

A divide; \$6 **C** multiply; \$42
B divide; \$7 **D** multiply; \$57

GO ON

Name ______________________________

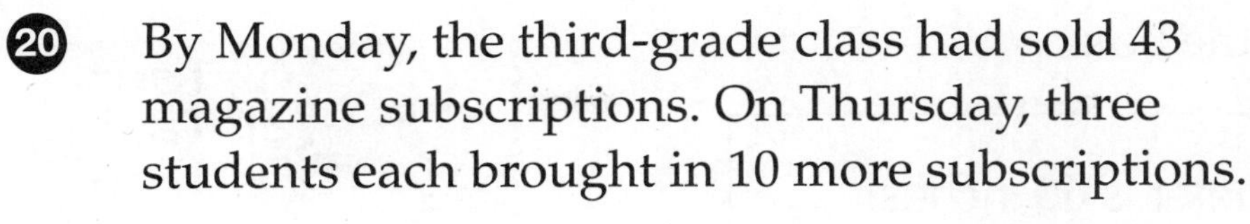

20 By Monday, the third-grade class had sold 43 magazine subscriptions. On Thursday, three students each brought in 10 more subscriptions.

How many subscriptions did the third graders sell in all?

On the lines below, explain how you solved the problem.

____

____

____

____

How many of each base-ten block can you use to model the problem?

21 Larry earned $8.00 for mowing the neighbor's lawn. He spent $4.95 on a collection of baseball cards. Does he have enough left over to buy 5 video game tokens at 50 cents each?

On the lines below, explain the steps you followed to solve this problem.

____

____

____

____

What do you need to figure out first?

Name ______________________________

 Read the riddle in the box.

My perimeter is 60 inches.
My sides are all equal.
Draw my shape.
How long is each side?

Illustrate and explain the answer to the riddle.

____
____
____
____
____

Can there be more than one answer?

---

23 Dina measured her book shelves. Each shelf is 3 feet long. There are 6 shelves in the bookcase.

Hector measured his book shelves. Each of his shelves is 24 inches long. There are 9 shelves in his bookcase.

Hector said that he had more space for books than Dina.

Is Hector right?

Remember 1 foot = 12 inches.

____
____
____

Test Taking Tips (Think · Solve · Explain — Short Answer)

What do you need to find out first?

Name ______________________________

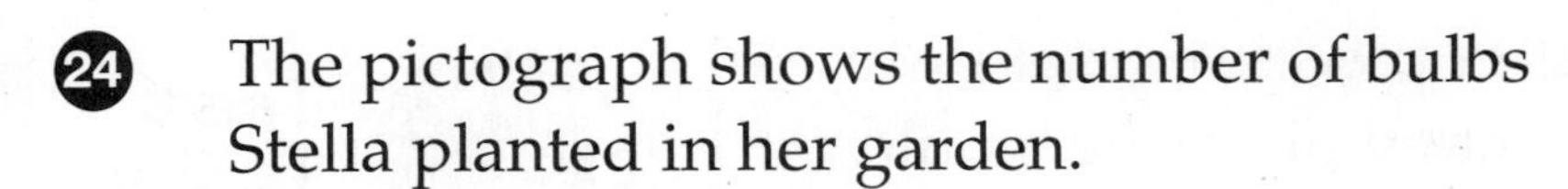

24 The pictograph shows the number of bulbs Stella planted in her garden.

Bulbs in Garden	
tulips	●●●●●●
daffodils	●●●●
crocuses	●●●●●●●●
Key: Each ● = 12 bulbs	

How many more crocuses did Stella plant than daffodils?

On the lines below, explain how you solved the problem.

____

____

Test Taking Tips

What do you need to find first?

25 An adult manatee weighs about 1,500 pounds. About how much would two adult manatees weigh?

Explain how you got your answer.

____

____

____

Test Taking Tips

How can using mental math help you solve the problem?

GO ON

Name ______________________________

26 Calvin picked up about 19 pieces of litter in 15 minutes. Estimate how many pieces he would pick up in one hour. ____________

Explain how you solved the problem.

____

____

____

Test Taking Tips

How many 15-minute periods are there in one hour? How can rounding help you solve the problem?

27 For every 5 cups of juice that Julia buys, she gets a ticket for a free pretzel. Julia has collected 14 tickets.

How many cups of juice has Julia bought?

____

Explain how you found your answer.

____

____

____

What operation can you use to find the answer?

GO ON

Name ____

28 **a.** Maggie bought 3 packs of cards. Each pack had 10 cards and 3 pieces of bubble gum.

If Maggie gave 5 cards to her brother, how many cards did she have left?

What information do you need to solve the problem?

GO ON

Name ______________________________

28 b. Write your answer on this page and explain your reasoning.

____

____

____

____

____

____

____

____

____

____

____

____

____

____

____

____

____

____

Test Taking Tips

How can making a model help you solve the problem?

GO ON

Name ____

29 **a.** There are 8 jars on the art room table. Half of them are holding 10 brushes each. The rest are holding 12 brushes each.

How many more brushes are needed if 100 students want to paint?

What do you need to figure out first?

Name ____________________________________

29 b. Write your answer on this page. Use words and pictures to make your answer clear.

____

____

____

____

____

____

____

____

____

____

____

____

____

____

____

Think • Solve • Explain
Long Answer

Test Taking Tips

Be sure that your explanation is clear and complete.

Answer Keys

Georgia Quality Core Curriculum Objectives

3.1 Applies estimation strategies beginning with front-end estimation and simple compatible numbers to predict appropriate results (see computation objectives).

3.19 Identifies even and odd numbers.

3.25 Rounds two- and three-digit numbers to the nearer ten or hundred.

3.30 Uses the terms: all, some, and none.

3.34 Identifies information needed to solve a given problem.

3.37 Solves one- and two-step word problems related to appropriate third-grade objectives. Includes oral and written problems and problem with extraneous information as well as information from sources such as pictographs, bar graphs, tables, and charts.

3.46 Uses properties of addition and multiplication (including commutative, associative, and properties of zero and one).

3.48 Adds and subtracts whole numbers (one-, two-, and three-digits, without or with regrouping), initially using manipulatives and then connecting the manipulations to symbolic procedures (problems presented vertically and horizontally with the horizontal problems rewritten vertically).

3.50 Applies mental computation strategies (such as counting up, counting back, simple compatible numbers, doubles, making ten, multiples of ten) to addition and subtraction, and to simple multiplication and division.

1. Answer: E NOT HERE

Discussion

Tip: If you solve the problem and don't see your solution listed, mark NOT HERE as the answer.

Students should begin each problem by trying to find the solution. If they can solve the problem and don't see the answer listed, then they can eliminate the first four answer choices. If they have trouble solving the problem, encourage them to test each answer choice to see whether or not it can be eliminated.

Item Numbers	Georgia QCC Objectives
1. E	3.48, 3.50
2. H	3.48
3. D	3.48
4. G	3.48
5. C	3.48
6. H	3.48
7. D	3.48
8. F	3.48
9. A	3.48
10. F	3.46
11. A	3.50
12. F	3.1, 3.25
13. D	3.19, 3.30, 3.37 3.48
14. H	3.37, 3.48
15. D	3.37, 3.48
16. H	3.34, 3.35
17. B	3.37, 3.48
18. G	3.37, 3.48
19. C	3.48

20. Answer: 7 socks

Discussion

Tip: What do you know about fact families that will help you solve the problem?

You can write a number sentence with a missing number, like this: 16 – ? = 9. Another number sentence that would work for this problem is 9 + ? = 16. Knowing the fact family for 7, 9, 16 will help you find the missing number.

3.34 Identifies information needed to solve a given problem.

3.35 Selects appropriate operation (addition, subtraction, or multiplication) for a given problem situation.

3.50 Applies mental computation strategies (such as counting up, counting back, simple compatible numbers, doubles, making ten, multiples of ten) to addition and subtraction, and to simple multiplication and division.

21. Answer: Tara can give Ellen 8 crayons and keep 7 for herself, or she can give Ellen 7 crayons and keep 8 for herself. She can also leave out one crayon and give each person 7 crayons.

Discussion

Tip: What do you know about fact families that will help you solve this problem?

You can make a list of fact families that include addends that equal 15. For example, you can make the family 6, 9, 15 or 7, 8, 15. The family 7, 8, 15 has the two addends that are closest to the same amount. So the best way for Tara to divide the 15 crayons nearly equally is in groups of 7 and 8.

Tip: How can you check your answers?

You can add the number of crayons each girl gets to make sure the total is 15.

3.34 Identifies information needed to solve a given problem.

3.35 Selects appropriate operation (addition, subtraction, or multiplication) for a given problem situation.

3.50Applies mental computation strategies (such as counting up, counting back, simple compatible numbers, doubles, making ten, multiples of ten) to addition and subtraction, and to simple multiplication and division.

22. Answer: Marie has about 60 holiday decorations.

Discussion

Tip: How can rounding up and rounding down help you solve the problem?

You can round 18 up to 20, and 22 down to 20. Then you can round 9 up to 10, and 12 down to 10. You can mentally add 20 + 20 + 10 + 10 to find that Marie has about 60 decorations.

Another strategy is to count on, starting with 20, and counting on 20 more, then 10, and then another 10.

3.25 Rounds two- and three-digit numbers to the nearer ten or hundred.

3.36 Employs problem-solving strategies (e.g., draw a picture; make a chart, graph, or table; guess and check; look for a pattern).

3.50 Applies mental computation strategies (such as counting up, counting back, simple compatible numbers, doubles, making ten, multiples of ten) to addition and subtraction, and to simple multiplication and division.

23. Answer: about 600

Discussion

Tip: Are you looking for an exact answer or an estimate?

You are looking for an estimate, so you can round 105 down to 100. Then you can subtract.

700 pencils – 100 pencils = 600 pencils

Another strategy is to round 105 down to 100 and then count on by 100's to 700. You will count on 6 times, so there is a difference of about 600.

3.25 Rounds two- and three-digit numbers to the nearer ten or hundred.

3.34 Identifies information needed to solve a given problem.

24. Answer: 22 students

Discussion

Tip: How can you use base-ten blocks to model the problem?

You can use 2 tens blocks and 9 ones blocks to model 29. Then subtract 7 to model the 7 students who can have acting roles. Count the number of blocks that are left to find the number of students who will not have acting roles.

Another strategy is to write a number sentence.

29 – 7 = 22.

3.35 Selects appropriate operation (addition, subtraction, or multiplication) for a given problem situation.

3.36 Employs problem-solving strategies (e.g., draw a picture; make a chart, graph, or table; guess and check; look for a pattern).

3.50 Applies mental computation strategies (such as counting up, counting back, simple compatible numbers, doubles, making ten, multiples of ten) to addition and subtraction, and to simple multiplication and division.

25. Answer: 8 km per hour faster

Discussion

Tip: What information do you need from the chart before you can solve this problem?

You need to know the speed of the zebra and the speed of the rabbit. The zebra can run 64 km per hour. The rabbit can run 56 km per hour.

You can subtract to compare the speeds.

64 – 56 = 8 km per hour faster

3.34 Identifies information needed to solve a given problem.

3.50 Applies mental computation strategies (such as counting up, counting back, simple compatible numbers, doubles, making ten, multiples of ten) to addition and subtraction, and to simple multiplication and division.

26. Answer: 24 dogs

Discussion

Tip: How can you restate the problem?

You could say: The vet treated 29 dogs on Monday. By the end of Tuesday, she had treated 53 dogs in all. How many dogs did she treat on Tuesday?

To solve the problem, you can write a number sentence with a missing addend: 29 + ? = 53. Then you can count on from 19 to 53 to find out that the vet treated 24 dogs on Tuesday.

Another possible strategy would be to write the problem as a subtraction number sentence: 53 – 19 = 24.

3.36 Employs problem-solving strategies (e.g., draw a picture; make a chart, graph, or table; guess and check; look for a pattern).

3.50 Applies mental computation strategies (such as counting up, counting back, simple compatible numbers, doubles, making ten, multiples of ten) to addition and subtraction, and to simple multiplication and division.

27. Answer: 352 meters

Discussion

Tip: What information from the map is useful in solving this problem?

To solve this problem, you need to find the distances from the library to the park, and then from the park to the post office. You don't need to consider the other distances that are shown on the map. To find the total distance Lisa walked, you would add 96 m (the distance from the library to the park) and 256 m (the distance from the park to the post office).

3.34 Identifies information needed to solve a given problem.

3.36 Employs problem-solving strategies (e.g., draw a picture; make a chart, graph, or table; guess and check; look for a pattern).

3.50 Applies mental computation strategies (such as counting up, counting back, simple compatible numbers, doubles, making ten, multiples of ten) to addition and subtraction, and to simple multiplication and division.

28. Answer: 4 pies

Discussion

Tip: How many pies have already been baked?

One strategy is to add the number already baked and subtract that total from 20. To find out how many pies have already been baked, you can write a number sentence: 8 pies (Paul) + 2 pies (first aunt) + 2 pies (second aunt) + 4 pies (mother) = 16 pies. Then you can subtract 16 from 20 to find that 4 pies are still needed.

Tip: How can drawing a picture help you solve the problem?

You can draw 20 circles to represent the 20 pies needed. Then label the circles P, A, and M to show how many Paul, his aunts, and his mother baked. This leaves 4 circles without a label. So you know that 4 pies still need to be baked.

3.35 Selects appropriate operation (addition, subtraction, or multiplication) for a given problem situation.

3.36 Employs problem-solving strategies (e.g., draw a picture; make a chart, graph, or table; guess and check; look for a pattern).

3.50 Applies mental computation strategies (such as counting up, counting back, simple compatible numbers, doubles, making ten, multiples of ten) to addition and subtraction, and to simple multiplication and division.

29. Answer: Two adults and two children.

Discussion

Tip: What information from the chart do you need?

The problem says that Pablo's children will go on the tour. To find out if only children went, you can count by 3's to see if you can stop at 20. Because you cannot stop exactly at 20, you know that adults also went on the tour.

Use guess-and-check. Suppose one adult went on the tour. Subtract to see how much money is left for children's tickets.

$20 – $7 = $13

Thirteen dollars would be left for children's tickets. But you can't skip count by 3's to exactly 13, so guess again.

Suppose two adults went on the tour. Find out how much two adult tickets cost and subtract that amount from $20 to see how much money is left for children's tickets.

$7 + $7 = $14

$20 – $14 = $6

Six dollars is enough for two children's tickets. So two adults and two children went on the tour.

3.34 Identifies information needed to solve a given problem.

3.36 Employs problem-solving strategies (e.g., draw a picture; make a chart, graph, or table; guess and check; look for a pattern).

3.50 Applies mental computation strategies (such as counting up, counting back, simple compatible numbers, doubles, making ten, multiples of ten) to addition and subtraction, and to simple multiplication and division.

Answer Keys

Georgia Quality Core Curriculum Objectives

3.11 Measures, using appropriate instruments and appropriate units, length, capacity, weight/mass, time, and temperature. Length: millimeter, inch, centimeter, foot, meter, yard, kilometer, mile. Capacity: milliliter, ounce, liter, cup, pint (liquid and dry), quart (liquid and dry), gallon. Weight/Mass: gram, ounce, kilogram, pound. Time: second, week, minute, month, hour, year, day, decade, century. Temperature: degree Fahrenheit, degree Celsius.

3.14 Determines and estimates amounts of money up to $5.00. Includes amounts spent, change received, and equivalent amounts.

3.15 Tells time to the minute and measures elapsed time, and measures time before and after the hour.

3.32 Continues or completes a given number sequence counting by ones, twos, threes, fours, fives, tens, hundreds, and thousands (include skip-counting on a number line).

3.37 Solves one- and two-step word problems related to appropriate third-grade objectives. Includes oral and written problems and problem with extraneous information as well as information from sources such as pictographs, bar graphs, tables, and charts.

3.40 Collects, reads, interprets, and compares data in charts, tables, and graphs.

3.50 Applies mental computation strategies (such as counting up, counting back, simple compatible numbers, doubles, making ten, multiples of ten) to addition and subtraction, and to simple multiplication and division.

1. Answer: D NOT HERE

Discussion

Tip: Which part tells the hour on a digital clock?

Students should be familiar with a digital clock. To review the meaning of the numbers (hours, minutes) and the ":", you could ask these questions:

What is to the left of the ":"?
What is to the right of the ":"?
Does a digital clock tell the minutes after or before the hour?

Have students work in groups. One student shows a time on a digital clock and the others write it down. They can take turns being the one who shows the time.

Item Numbers	Georgia QCC Objectives
1. D	3.15
2. H	3.15
3. C	3.11, 3.32
4. F	3.15, 3.37
5. D	3.15, 3.37
6. G	3.11, 3.37
7. C	3.11, 3.40
8. G	3.40
9. A	3.15
10. J	3.14
11. D	3.50
12. H	3.50
13. A	3.37, 3.50
14. G	3.11, 3.40
15. A	3.11, 3.40

16. Answer: See discussion below.

Discussion

Tip: How can counting by 15's help you find a pattern?

The pattern is that the lesson is scheduled for 15 minutes later each day. Counting by 15's helps you fill in the missing times.

Guitar Lessons	
Monday	12:30
Tuesday	12:45
Wednesday	**1:00**
Thursday	1:15
Friday	1:30
Saturday	**1:45**

3.31 Determines a pair of numbers or the missing element of a pair when given a relation or rule. Determines the relation or rule when given pairs of numbers.

3.34 Identifies information needed to solve a given problem.

3.36 Employs problem-solving strategies (e.g., draw a picture; make a chart, graph, or table; guess and check; look for a pattern).

3.50 Applies mental computation strategies (such as counting up, counting back, simple compatible numbers, doubles, making ten, multiples of ten) to addition and subtraction, and to simple multiplication and division.

17. Answer: 12 hours and 18 minutes

Discussion

Tip: What are you trying to find out? Restate the problem in your own words.

You are trying to find out how much time passed between 6:30 A.M. and 6:48 P.M. You know that from 6:30 to 6:30 is 12 hours. Then you can count on from 6:30 to 6:48 to find that 18 more minutes passed.

Another strategy is to act it out, using a clock with movable hands.

3.35 Selects appropriate operation (addition, subtraction, or multiplication) for a given problem situation.

3.36 Employs problem-solving strategies (e.g., draw a picture; make a chart, graph, or table; guess and check; look for a pattern).

3.50 Applies mental computation strategies (such as counting up, counting back, simple compatible numbers, doubles, making ten, multiples of ten) to addition and subtraction, and to simple multiplication and division.

18. Answer: 3 minutes

Discussion

Tip: How can finding each stop time help you solve the problem?

You can make a list of ending times, based on the starting times given. Add 30 minutes to each time, since that is the time limit. Dwaine will have to stop at 4:38, Suzanne will have to stop at 4:20, Oscar will have to stop at 4:45, and Loretta will have to stop at 4:29. Since it is now 4:17, Cheryl has to wait 3 minutes for Suzanne to stop.

3.15 Tells time to the minute and measures elapsed time, and measures time before and after the hour.

3.34 Identifies information needed to solve a given problem.

3.36 Employs problem-solving strategies (e.g., draw a picture; make a chart, graph, or table; guess and check; look for a pattern).

3.50 Applies mental computation strategies (such as counting up, counting back, simple compatible numbers, doubles, making ten, multiples of ten) to addition and subtraction, and to simple multiplication and division.

19. Answer: $2.77. Possible combination of bills and coins: two one-dollar bills, three quarters, two pennies.

Discussion

Tip: How can acting it out help you solve this problem?

You can use real money or play money to act out this problem. Start with $17.23 and add up to $20.00 by counting out two pennies as you count up to $17.25. Then count out three quarters as you count up to $18.00. Then count out two dollar bills as you count up to $20.00.

Another strategy is to write a subtraction problem and solve.

$$\begin{array}{r} \$20.00 \\ -\ 17.23 \\ \hline \$\ 2.77 \end{array}$$

3.35 Selects appropriate operation (addition, subtraction, or multiplication) for a given problem situation.

3.36 Employs problem-solving strategies (e.g., draw a picture; make a chart, graph, or table; guess and check; look for a pattern).

20. Answer: 6 times

Discussion

Tip: What operation can you use to solve the problem?

You can guess and check. If you guess 4, then you would add $1.50 four times to see if it adds up to enough money. You would find that 4 is not enough times, so you would change your guess. If you guess 6 times, you would add $1.50 six times. It adds up to $9.00, which is enough money to buy a circus ticket.

Another strategy is to add $1.50 over and over until you get close to $8.75.

3.34 Identifies information needed to solve a given problem.

3.35 Selects appropriate operation (addition, subtraction, or multiplication) for a given problem situation.

3.36 Employs problem-solving strategies (e.g., draw a picture; make a chart, graph, or table; guess and check; look for a pattern).

21. Answer: The least expensive item is the food cooler, and the most expensive item is the backpack.

Discussion

Tip: What do you know about ordering numbers that will help you solve this problem?

Since each price has a different dollar amount, you can ignore the cents as you put the amounts in order from lowest to highest. The price of $9 is the highest one on the list, and the price of $6 is the lowest.

3.28 Compares and orders whole numbers through 9,999.

3.34 Identifies information needed to solve a given problem.

3.36 Employs problem-solving strategies (e.g., draw a picture; make a chart, graph, or table; guess and check; look for a pattern).

22. Answer: $45.00

Discussion

Tip: How can rounding to the nearest dollar help you solve the problem?

You can round $19.95 up to $20, $14.96 up to $15.00, and $4.95 up to $5.00. Since Julio chose 2 pairs of socks, you can double the $5.00 to get $10.00. Then add $20.00 + $15.00 + $10.00.

3.34 Identifies information needed to solve a given problem.

3.35 Selects appropriate operation (addition, subtraction, or multiplication) for a given problem situation.

3.50 Applies mental computation strategies (such as counting up, counting back, simple compatible numbers, doubles, making ten, multiples of ten) to addition and subtraction, and to simple multiplication and division.

23. Answer: Yes; see discussion below.

Discussion

Tip: What are you trying to find out? Restate the problem in your own words.

You can restate the problem in this way: Is $20 more than or less than the sum of $11.88, $4.06, and $3.30?
To solve the problem, you can round to the nearest dollar. $11.88 rounds up to $12.00. $4.06 rounds down to $4.00. $3.30 rounds down to $3.00. If you add $12.00 + $4.00 + $3.00, you get $19.00, which is less than $20.00, so Anthony has enough money to buy the three books.
Another strategy is to subtract the rounded numbers from 20, to see if you have any left.

$20.00 - $12.00 = $8.00
$8.00 - $4.00 = $4.00.
$4.00 - $3.00 = $1.00.

There is $1.00 left over, so Anthony has enough money.

3.34 Identifies information needed to solve a given problem.

3.35 Selects appropriate operation (addition, subtraction, or multiplication) for a given problem situation.

3.36 Employs problem-solving strategies (e.g., draw a picture; make a chart, graph, or table; guess and check; look for a pattern).

24. Answer: See discussion below.

Discussion

Tip: How can drawing a picture help you solve the problem?

You can use a clock with movable hands to figure out the times. The time Kristin looked at the clock is given, so you can move the hands to show 5:30. Move the minute hand 30 minutes ahead, and you can see the party will be over at 6:00. Starting at 5:30, move the hour hand back three hours, and you can see the party started at 2:30.

Party Started

Kristin Looked at Clock

Party Will End

Tip: How can rereading the question help you check your answer?

As you reread the question, point to each clock as the time is mentioned. Ask yourself if the time you have shown matches the time mentioned in the question.

3.15 Tells time to the minute and measures elapsed time, and measures time before and after the hour.

3.36 Employs problem-solving strategies (e.g., draw a picture; make a chart, graph, or table; guess and check; look for a pattern).

3.50 Applies mental computation strategies (such as counting up, counting back, simple compatible numbers, doubles, making ten, multiples of ten) to addition and subtraction, and to simple multiplication and division.

25. Answer: Brian has a total of $4.03 in change. This is not enough to buy a baseball cap for $4.95. He needs $.92 more.

Discussion

Tip: How can making a list help you find the answer?

You can make a list of the money amounts for each category of coin. Brian has $1.50 in half dollars, $1.25 in quarters, $.70 in dimes, $.55 in nickels, and $.03 in pennies. If you write these amounts in a vertical list, you can add them. Once you have added the amounts, you can compare the total to $4.95, the price of the baseball cap. Then you can see that it is not enough.

Another strategy is to use a calculator or mental mathematics.

Tip: How can you be sure that your explanation is clear and complete.

You can check that you have added all the amounts correctly and have compared the total to the price of the cap.

3.14 Determines and estimates amounts of money up to $5.00. includes amounts spent, change received, and equivalent amounts.

3.36 Employs problem-solving strategies (e.g., draw a picture; make a chart, graph, or table; guess and check; look for a pattern).

3.40 Collects, reads, interprets, and compares data in charts, tables, and graphs.

3.50 Applies mental computation strategies (such as counting up, counting back, simple compatible numbers, doubles, making ten, multiples of ten) to addition and subtraction, and to simple multiplication and division.

Answer Keys

Georgia Quality Core Curriculum Objectives

3.1 Applies estimation strategies beginning with front-end estimation and simple compatible numbers to predict appropriate results (see computation objectives).

3.11 Measures, using appropriate instruments and appropriate units, length, capacity, weight/mass, time, and temperature. Length: Millimeter, Inch, Centimeter, Foot, Meter, Yard, Kilometer, Mile. Capacity: Milliliter, Ounce, Liter, Cup, Pint (Liquid and Dry), Quart (Liquid and Dry), Gallon. Weight/Mass: Gram, Ounce, Kilogram, Pound. Time: Second, Week, Minute, Month, Hour, Year, Day Decade, Century. Temperature: Degree, Fahrenheit, Degree Celsius.

3.14 Determines and estimates amounts of money up to $5.00. includes amounts spent, change received, and equivalent amounts.

3.17 Translates words to numerals and numerals to words up to 9,999.

3.18 Recognizes different names for whole numbers through 9,999 including names in expanded notation form (9,000 + 900 + 90 + 9; 9 thousand, 9 hundreds, 9 tens, 9 ones; nine thousand, nine hundred, ninety-nine).

3.20 Determines ordinal numbers through twentieth.

3.21 Relates concrete and pictorial models to numbers through thousands, and relates numbers to models; names numbers orally.

3.22 Identifies place value through thousands and identifies the number of thousands, hundreds, tens, and ones in a given number.

3.23 Identifies the rational number (whole numbers and simple fractions) corresponding to a given point on the number line.

3.24 Uses a number line to determine to which multiple of 10 or 100 a given number (up to 1,000) is nearer.

3.25 Rounds two- and three-digit numbers to the nearer ten or hundred.

3.26 Recognizes numerical relationships through 9,999 (such as between, before, after, equal to, nearest to, least, and greatest).

3.28 Compares and orders whole numbers through 9,999.

3.31 Determines a pair of numbers or the missing element of a pair when given a relation or rule. Determines the relation or rule when given pairs of numbers.

3.36 Employs problem-solving strategies (e.g., draw a picture; make a chart, graph, or table; guess and check; look for a pattern).

3.40 Collects, reads, interprets, and compares data in charts, tables, and graphs.

3.48 Adds and subtracts whole numbers (one-, two-, and three-digits, without or with regrouping), initially using manipulatives and then connecting the manipulations to symbolic procedures (problems presented vertically and horizontally with the horizontal problems rewritten vertically).

1. Answer: A 10

Discussion

Tip: How could you use skip-counting to find the answer?

Students should be familiar with base-ten blocks. Review the place value that these blocks represent by asking these questions:

Which block shows the ones place? the tens place?
How many ones blocks are in a tens block?
Which block shows the hundreds place?
How many tens blocks are in a hundreds block?

Have students divide into pairs and practice making these numbers using base-ten blocks: 346, 587, and 241.

Item Numbers	Georgia QCC Objectives
1. A	3.22
2. G	3.14
3. D	3.11, 3.20, 3.40
4. H	3.11, 3.40
5. C	3.22
6. J	3.31, 3.36, 3.48
7. D	3.31, 3.36, 3.48
8. G	3.14, 3.36
9. A	3.31, 3.36, 3.48
10. H	3.31, 3.36, 3.48
11. B	3.17
12. G	3.18
13. C	3.36, 3.40
14. G	3.1, 3.36, 3.40
15. D	3.26, 3.28
16. H	3.23, 3.24, 3.26
17. D	3.25
18. H	3.21, 3.26

19. Answer: Four hundreds has the value of 400, zero tens has the value of 0, and 5 ones has the value of 5.

Discussion

Tip: How can you use place value to answer the question?

You can model the number 405 with base-ten blocks, using 4 base-ten hundreds, and 5 base-ten ones.

Another strategy is to use a place-value chart:

Hundreds	Tens	Ones
4	0	5

3.22 Identifies place value through thousands and identifies the number of thousands, hundreds, tens, and ones in a given number.

3.28 Compares and orders whole numbers through 9,999.

3.36 Employs problem-solving strategies (e.g., draw a picture; make a chart, graph, or table; guess and check; look for a pattern).

20. Answer: 43 students

Discussion

Tip: What information in the chart can help you solve the problem?

How Third Graders Get to School	
Ride the bus	18
Walk	12
Ride bikes	14
Ride in a car	11

You can add the numbers of students who ride the bus, ride bikes, and ride in cars.

You can use a number sentence to show how you solved the problem.

18 students + 14 students + 11 students = 43 students

3.34 Identifies information needed to solve a given problem.

3.35 Selects appropriate operation (addition, subtraction, or multiplication) for a given problem situation.

3.40 Collects, reads, interprets, and compares data in charts, tables, and graphs.

3.50 Applies mental computation strategies (such as counting up, counting back, simple compatible numbers, doubles, making ten, multiples of ten) to addition and subtraction, and to simple multiplication and division.

21. Answer: 200 miles

Discussion

Tip: How can using a place-value chart help you solve this problem?

Look at the position of the numeral 5. It is in the hundreds place. You can use a place-value chart to show the change.

Thousands	Hundreds	Tens	Ones
6	5	9	2

The value of the 5 is 500. If the 5 changes to a 7, its value would be 700. To find the difference in values, you can subtract.

700 – 500 = 200. So Mr. Langley has driven 200 miles.

3.22 Identifies place value through thousands and identifies the number of thousands, hundreds, tens, and ones in a given number.

3.34 Identifies information needed to solve a given problem.

3.35 Selects appropriate operation (addition, subtraction, or multiplication) for a given problem situation.

22. Answer: Lori

Discussion

Tip: What do you know about place value that will help you solve this problem?

You can use place value to decide which number has greater value. Think about how many tens each number has. The number 53 has 5 tens. The number 35 has only 3 tens. You know that 5 tens is more than 3 tens so 53 is more than 35.

3.22 Identifies place value through thousands and identifies the number of thousands, hundreds, tens, and ones in a given number.

3.28 Compares and orders whole numbers through 9,999.

3.36 Employs problem-solving strategies (e.g., draw a picture; make a chart, graph, or table; guess and check; look for a pattern).

23. Answer: 384, 394, 404, 414

Discussion

Tip: How can finding a pattern help you solve the problem?

You can look for a pattern and see that each number is 10 more than the one before it. To continue the pattern, you add 10 to each number.

3.22 Identifies place value through thousands and identifies the number of thousands, hundreds, tens, and ones in a given number.

3.31 Determines a pair of numbers or the missing element of a pair when given a relation or rule. Determines the relation or rule when given pairs of numbers.

3.36 Employs problem-solving strategies (e.g., draw a picture; make a chart, graph, or table; guess and check; look for a pattern).

24. Answer: Yes, an estimate of 250 pounds is reasonable.

Discussion

Tip: When do you round up? When do you round down?

If the number in the ones place is 1 to 4, you round down. If the number in the ones place is 5 to 9, you round up. You can round 57 up to 60, 43 down to 40, 68 up to 70, 32 down to 30, and 47 up to 50. If you add the rounded numbers, you can see that an estimate of 250 is reasonable.

3.25 Rounds two- and three-digit numbers to the nearer ten or hundred.

3.36 Employs problem-solving strategies (e.g., draw a picture; make a chart, graph, or table; guess and check; look for a pattern).

3.50 Applies mental computation strategies (such as counting up, counting back, simple compatible numbers, doubles, making ten, multiples of ten) to addition and subtraction, and to simple multiplication and division.

25. Answer: The closest town is Swanee, and the farthest one is Burnic.

Discussion

Tip: How can making a list help you solve the problem?

You can make a list of the number of miles, in order from least distance to greatest distance.

Swanee	157
Sterling	176
Burnic	261

Then you can see which town is closest and which is farthest.

3.28 Compares and orders whole numbers through 9,999.

3.35 Selects appropriate operation (addition, subtraction, or multiplication) for a given problem situation.

3.36 Employs problem-solving strategies (e.g., draw a picture; make a chart, graph, or table; guess and check; look for a pattern).

26. Answer: 767

Discussion

Tip: How can ordering the numbers help you solve the problem?

If you order the numbers, you can see which number is the greatest in value. The list can be from least to greatest or greatest to least.

Another strategy is to locate each number on a number line.

3.28 Compares and orders whole numbers through 9,999.

3.35 Selects appropriate operation (addition, subtraction, or multiplication) for a given problem situation.

3.36 Employs problem-solving strategies (e.g., draw a picture; make a chart, graph, or table; guess and check; look for a pattern).

27. Answer: See discussion below.

Discussion

Tip: How can finding a number pattern in each column help you complete the table?

The pattern already established is counting by 2's in column 1, and counting by 50 cents in column 2. To complete the chart, you simply continue the pattern, filling in the blanks.

Tip: How can you check that your solution makes sense?

You can read down each column to be sure the pattern is continued.

The missing parts of the chart should be filled in as follows.

Students should also add three more categories, either at the beginning or at the end, as follows: 2, $.50; 4, $1.00; 6, $1.50; 24, $6.00; 26, $6.50; 28, $7.00.

Number of Tickets Won	Prize Value
8	$2.00
10	$2.50
12	$3.00
14	$3.50
16	$4.00
18	$4.50
20	$5.00
22	$5.50

3.31 Determines a pair of numbers or the missing element of a pair when given a relation or rule. Determines the relation or rule when given pairs of numbers.

3.34 Identifies information needed to solve a given problem.

3.36 Employs problem-solving strategies (e.g., draw a picture; make a chart, graph, or table; guess and check; look for a pattern).

3.40 Collects, reads, interprets, and compares data in charts, tables, and graphs.

28. Answer: Accept any answer between 150 and 170.

Discussion

Tip: About how many signs does Koko learn every 2 months?

By counting on or subtracting, you can see that Koko learns about 20 signs every 2 months. If you add 20 to the amount Koko knew at age 44 months, you would get 155. Allowing for variation, a reasonable amount at age 46 months would be between 150 and 170.

3.31 Determines a pair of numbers or the missing element of a pair when given a relation or rule. Determines the relation or rule when given pairs of numbers.

3.34 Identifies information needed to solve a given problem.

3.40 Collects, reads, interprets, and compares data in charts, tables, and graphs.

Answer Keys

Georgia Quality Core Curriculum Objectives

3.23 Identifies the rational number (whole numbers and simple fractions) corresponding to a given point on the number line.

3.27 Writes a number sentence represented by a picture or an array.

3.30 Uses the terms: all, some, and none.

3.34 Identifies information needed to solve a given problem.

3.36 Employs problem-solving strategies (e.g., draw a picture; make a chart, graph, or table; guess and check; look for a pattern).

3.37 Solves one- and two-step word problems related to appropriate third grade objectives. Includes oral and written problems and problem with extraneous information as well as information from sources such as pictographs, bar graphs, tables, and charts.

3.42 Multiples whole numbers up to two-digit by one-digit numbers using models and three-digit by one-digit numbers using computational strategies.

3.43 Relates concrete and pictorial models to multiplication and division.

3.44 Relates division to multiplication and uses models such as partitioning, and repeated subtraction to divide one- and two-digit numbers by one-digit numbers without and with remainders.

3.45 Determines basic multiplication and division facts through 9 x 9 by using strategies such as skip-counting, multiplying by zero and one, dividing by one, splitting arrays, commutative property of multiplication, and using known facts to find unknown facts.

3.47 Selects appropriate symbol (+, –, x, —, <, >, =) for use in a number sentence.

3.49 Recalls basic multiplication facts through 9 x 9.

1. Answer: C 8

Discussion

Tip: Use what you know about multiplying by 1 to solve the problem.

Students should recognize that when you multiply a number by 1, the product is always that number. Ask questions such as these to make sure students remember this.

> What is 1×7? 1×9? 1×12?
> What do you notice about each number when you multiply it by 1?

Some students may confuse addition and multiplication by 1, thinking that the product is always 1 more than the number. Reinforcing the Identity Property for Multiplication with examples will help these students.

Item Numbers	Georgia QCC Objectives
1. C	3.45, 3.49
2. J	3.49
3. D	3.37, 3.49
4. H	3.30, 3.37, 3.45
5. B	3.45, 3.49
6. J	3.45, 3.49
7. B	3.45
8. F	3.45
9. D	3.45
10. J	3.45
11. A	3.30, 3.37, 3.45
12. F	3.27, 3.42, 3.43, 3.45
13. C	3.37
14. G	3.44
15. D	3.34, 3.36, 3.37
16. F	3.36, 3.37
17. D	3.23, 3.42, 3.43
18. G	3.44, 3.49
19. B	3.43
20. G	3.37, 3.45

21. Answer: 16

Discussion

Tip: How can using a multiplication table help you solve this problem?

You can use a multiplication table to find out how much 4 times 4 is.

Other strategies: You can skip count by 4's four times or you can add 4 four times.

$4 + 4 + 4 + 4 = 16$

3.36 Employs problem-solving strategies (e.g., draw a picture; make a chart, graph, or table; guess and check; look for a pattern).

3.50 Applies mental computation strategies (such as counting up, counting back, simple compatible numbers, doubles, making ten, multiples of ten) to addition and subtraction, and to simple multiplication and division.

22. Answer: 24 muffins

Discussion

Tip: How can drawing a picture help solve the problem? What operation can you use to solve the problem?

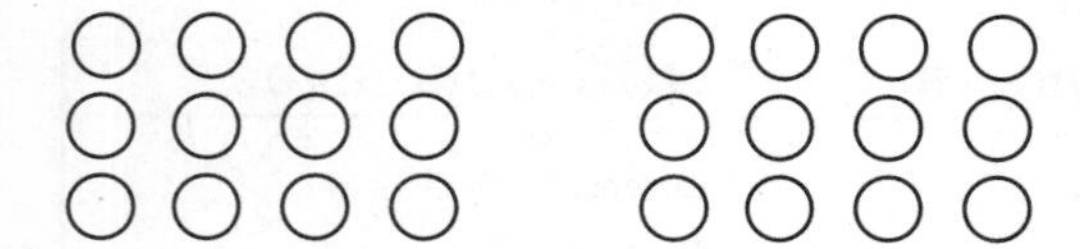

You can draw a picture to show the information. Each row represents the number of muffins for one family member. You can count the muffins to find the total.

Another strategy is to multiply the number of family members by the number of muffins.

$6 \times 4 = 24$ muffins

3.35 Selects appropriate operation (addition, subtraction, or multiplication) for a given problem situation.

3.36 Employs problem-solving strategies (e.g., draw a picture; make a chart, graph, or table; guess and check; look for a pattern).

3.50 Applies mental computation strategies (such as counting up, counting back, simple compatible numbers, doubles, making ten, multiples of ten) to addition and subtraction, and to simple multiplication and division.

23. Answer: C (35 seconds)

Discussion

Tip: What do you need to figure out first?

First, you have to figure out how many floors the elevator will be going. You can count on or subtract to find that it will go 5 floors. The next step is to multiply 5 floors by 7 seconds to get 35 seconds.

Another strategy is to use repeated addition.

$7 + 7 + 7 + 7 + 7 = 35$ seconds

3.35 Selects appropriate operation (addition, subtraction, or multiplication) for a given problem situation.

3.50 Applies mental computation strategies (such as counting up, counting back, simple compatible numbers, doubles, making ten, multiples of ten) to addition and subtraction, and to simple multiplication and division.

24. Answer: 36, 42, 48

Discussion

Tip: How is each number related to the number that comes before it?

The numbers are multiples of 6. Remember your multiplication table: 6 is 1 x 6; 12 is 2 x 6; and so forth. You can multiply or count by sixes to find the next three numbers.

3.31 Determines a pair of numbers or the missing element of a pair when given a relation or rule. Determines the relation or rule when given pairs of numbers.

3.50 Applies mental computation strategies (such as counting up, counting back, simple compatible numbers, doubles, making ten, multiples of ten) to addition and subtraction, and to simple multiplication and division.

25. Answer: 36 players

Discussion

Tip: What operation can you use to solve the problem?

You can make a list:

Team 1	9 players
Team 2	9 players
Team 3	9 players
Team 4	9 players

Then you can add 9 + 9 + 9 + 9, or you can multiply 9 x 4.

3.35 Selects appropriate operation (addition, subtraction, or multiplication) for a given problem situation.

3.36 Employs problem-solving strategies (e.g., draw a picture; make a chart, graph, or table; guess and check; look for a pattern).

3.50 Applies mental computation strategies (such as counting up, counting back, simple compatible numbers, doubles, making ten, multiples of ten) to addition and subtraction, and to simple multiplication and division.

26. Answer: 9 nickels

Discussion

Tip: How could counting by fives help you solve the problem? Could you use an operation?

You could count by fives to 45 to find out how many nickels are in 45¢; or you can divide: 45¢ ÷ 5¢ = 9 nickels.

3.31 Determines a pair of numbers or the missing element of a pair when given a relation or rule. Determines the relation or rule when given pairs of numbers.

3.36 Employs problem-solving strategies (e.g., draw a picture; make a chart, graph, or table; guess and check; look for a pattern).

3.50 Applies mental computation strategies (such as counting up, counting back, simple compatible numbers, doubles, making ten, multiples of ten) to addition and subtraction, and to simple multiplication and division.

27. Answer: 9 rows

Discussion

Tip: How can making an array help you solve this problem?

You can make an array that starts with a row of 8. Then you can keep adding rows of 8 until the total is 72. After that, you can just count the rows.

Another strategy is to write a number sentence with a missing factor: 8 x ? = 72. Then you can guess and check until you find the missing factor.

3.35 Selects appropriate operation (addition, subtraction, or multiplication) for a given problem situation.

3.36 Employs problem-solving strategies (e.g., draw a picture; make a chart, graph, or table; guess and check; look for a pattern).

3.50 Applies mental computation strategies (such as counting up, counting back, simple compatible numbers, doubles, making ten, multiples of ten) to addition and subtraction, and to simple multiplication and division.

28. Answer: Division key

Discussion

Tip: How does the answer in the display compare to the numbers on the left side of the equation?

Since the answer, 9, is less than 72, you know to rule out the addition and multiplication keys. You know that subtracting 8 from 72 would yield an answer in the 60's. The division key seems to be the right choice, so you verify your choice: $72 \div 8 = 9$; $9 \times 8 = 72$.

3.36 Employs problem-solving strategies (e.g., draw a picture; make a chart, graph, or table; guess and check; look for a pattern).

3.50 Applies mental computation strategies (such as counting up, counting back, simple compatible numbers, doubles, making ten, multiples of ten) to addition and subtraction, and to simple multiplication and division.

29. Answers: 1. about $4.00, 2. about $8.00, 3. about $12.00, 4. about $7.00

Discussion

Tip: How can rounding to the nearest dollar help you find the answers?

You can use rounding to estimate with easier amounts.

$2.15 to $2.00

$1.99 to $2.00

$2.75 to $3.00

Then you can use mental math to multiply these rounded amounts by the number of items to find an estimated total for each item purchased. When more than two types of items are bought, you can use mental math to add the estimated totals.

Another strategy is to make a list of the items and the estimated amount for each item. Then add the amounts on each list.

You can also use a calculator to add the estimated amounts for each combination of items.

Tip: How can looking at a multiplication table help?

By looking at a multiplication table, you can check your answers for accuracy.

3.1 Applies estimation strategies beginning with front-end estimation and simple compatible numbers to predict appropriate results (see computation objectives).

3.35 Selects appropriate operation (addition, subtraction, or multiplication) for a given problem situation.

3.36 Employs problem-solving strategies (e.g., draw a picture; make a chart, graph, or table; guess and check; look for a pattern).

3.50 Applies mental computation strategies (such as counting up, counting back, simple compatible numbers, doubles, making ten, multiples of ten) to addition and subtraction, and to simple multiplication and division.

30. Answer: See table and discussion below.

Discussion

Tip: How can you use the picture to help solve the problem? How can you make a list to solve the problem?

You can circle equal groups of erasers to find out how many packages Jolene could use. You might start by lightly circling groups of two to check for leftovers and to count the number of groups. Then circle groups of three, and so on.

Without the picture you can choose a number of erasers for each package and then divide or skip count to check for leftovers.

Making a list helps you know how much work you have done and what steps you need to take.

To figure out how much to charge for a package, you can multiply the number of erasers in a package by 5¢. However, if you decided to put 12 erasers into 1 package, you may prefer to skip count by fives or break down the problem:

(6 erasers x 5¢) + (6 erasers x 5¢) =

30¢ + 30¢ = 60¢.

Number of Packages	Number of Erasers in One Package	Price per Package
1	12	60¢
2	6	30¢
3	4	20¢
4	3	15¢
6	2	10¢
12	1	5¢

3.14 Determines and estimates amounts of money up to $5.00. Includes amounts spent, change received, and equivalent amounts.

3.36 Employs problem-solving strategies (e.g., draw a picture; make a chart, graph, or table; guess and check; look for a pattern).

3.40 Collects, reads, interprets, and compares data in charts, tables, and graphs.

3.50 Applies mental computation strategies (such as counting up, counting back, simple compatible numbers, doubles, making ten, multiples of ten) to addition and subtraction, and to simple multiplication and division.

Answer Keys

Georgia Quality Core Curriculum Objectives

3.29 Identifies subsets of given sets.

3.34 Identifies information needed to solve a given problem.

3.39 Organizes data into charts and tables and constructs bar graphs using scales of one, two, five, or ten units and pictographs using scales of one, two, three, four, five, or ten units.

3.40 Collects, reads, interprets, and compares data in charts, tables, and graphs.

3.41 Determines probability of a given event through exploration (equally like, least likely, and most likely).

1. Answer: B 9 students

Discussion

Tip: Use the key to find out how many votes each wheel stands for.

Students may look at the pictograph and think that the answer is 3 because they do not notice the key at the bottom. Use questions such as this to show the significance of the key.

What does the key tell you about the number of votes?
How can you use the key to find the number of votes?
Which two operations could you use?

Discuss with the class the importance of looking at the entire graph carefully before trying to interpret it.

Item Numbers	Georgia QCC Objectives
1. B	3.39, 3.40
2. H	3.39, 3.40
3. C	3.29, 3.40
4. G	3.40
5. B	3.29, 3.40
6. J	3.40
7. C	3.40
8. F	3.40
9. D	3.39, 3.40
10. G	3.39, 3.40
11. B	3.39, 3.40
12. F	3.41
13. B	3.41
14. J	3.41
15. B	3.34, 3.41
16. J	3.41

17. Answer: See discussion below.

Discussion

Tip: How can counting by 5's help you check your work?

After you make four tally marks, you make the fifth one by drawing a line across the group of four. Then you can count the groups by 5's, counting on for any marks that are left over. Compare the final count to the number in the paragraph to make sure they match.

Third Graders' Favorite Picnic Food	
Sandwiches	卌 卌 卌 |
Chicken	卌 卌 卌 卌 ||||
Hamburgers	卌 ||||
Fruit Salad	卌 卌 ||

3.40 Collects, reads, interprets, and compares data in charts, tables, and graphs.

---

18. Answer: It is most likely that Jana will like dogs the best, followed by cats.

Discussion

Tip: How can making a list help you solve the problem?

To predict Jana's most likely choice, you need to know the most popular pet of the third graders. Put the numbers in a list from greatest to least. Then label the numbers with the correct pets. This makes it easy to see which is the most popular pet.

3.34 Identifies information needed to solve a given problem.

3.40 Collects, reads, interprets, and compares data in charts, tables, and graphs.

---

19. Answer: Possible choices are Goodies, Ring-a-Ling and Crispy. They are priced in the middle and received the most votes.

Discussion

Tip: Is there more than one right answer?

Yes. You can choose the chips based on number of votes alone. In this case, you would choose Goodies, Ring-a-Ling, and Crispy. You might choose to order the ones that cost the least. In this case, you would choose Toasties, Wavy, and Goodies.

3.34 Identifies information needed to solve a given problem.

3.40 Collects, reads, interprets, and compares data in charts, tables, and graphs.

20. Answer: See discussion below.

3.34 Identifies information needed to solve a given problem.

3.39 Organizes data into charts and tables and constructs bar graphs using scales of one, two, five, or ten units and pictographs using scales of one, two, three, four, five, or ten units.

Discussion

Tip: What does each box on the bar graph stand for?

Each box on the bar graph stands for one day, just as each tally mark does. If there are 4 tally marks representing 4 cloudy days, then you should fill in 4 boxes on the bar graph to represent 4 cloudy days.

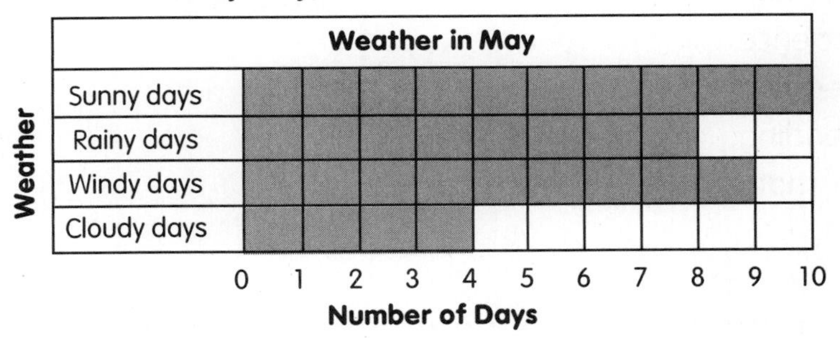

21. Answer: See discussion below.

3.34 Identifies information needed to solve a given problem.

3.41 Determines probability of a given event through exploration (equally like, least likely, and most likely).

Discussion

Tip: How can making a list of the marbles help you solve the problem?

Combination	Possible	Impossible
2 blue, 1 red, 1 black	X	
4 yellow	X	
1 green, 3 black		X
4 black		X

If you draw a picture of the different-colored marbles, you can see which possibilities there are. Because there are no green marbles, the third combination in the chart is not possible. Because there are only 3 black marbles, the fourth combination is not possible.

22. Answer: Nathan has 6 choices.

Discussion

Tip: How can making an organized list help you?

You can make an organized list like this to show Nathan's choices:

reading, math, science
reading, science, math
math, reading, science
math, science, reading
science, math, reading
science, reading, math

Then you can count the choices.

3.36 Employs problem-solving strategies (e.g., draw a picture; make a chart, graph, or table; guess and check; look for a pattern).

3.40 Collects, reads, interprets, and compares data in charts, tables, and graphs.

3.41 Determines probability of a given event through exploration (equally like, least likely, and most likely).

23. Answer: 6 ways

Discussion

Tip: How can making an organized list help you solve this problem?

An organized list like this one helps you see that there are 6 ways they can team up:

Ming-lo, Betty
Ming-lo, Jamal
Ming-lo, Sara
Betty, Jamal
Betty, Sara
Jamal, Sara

3.36 Employs problem-solving strategies (e.g., draw a picture; make a chart, graph, or table; guess and check; look for a pattern).

3.40 Collects, reads, interprets, and compares data in charts, tables, and graphs.

3.41 Determines probability of a given event through exploration (equally like, least likely, and most likely).

24. Answers: 1. Red, Blue, Green, and Yellow;
2. Yes, because the four sections are equal sizes;
3. 1 in 4;
4. 10 times for each color.

Discussion

Tip: How many sections are on the spinner? Are all sections the same size?

There are four sections on the spinner, and all sections are the same size. Because there are four different choices, and each one has the same amount of space on the spinner, all have an equal chance of being spun.

3.34 Identifies information needed to solve a given problem.

3.41 Determines probability of a given event through exploration (equally like, least likely, and most likely).

25. Answer: See discussion below.

Discussion

Tip: What information do you need to complete the graph?

To show the scores for Teams D, E, and F, you first have to figure out what their scores were. If Team D scored 5 more points than Team B, and Team B scored 50, you can add 5 to 50 to find Team D's score. If Team E scored 20 points more than Team C, and Team C scored 20, you can add 20 to 20 to find Team E's score. If Team F scored 5 points less than Team A, and Team A scored 30, you can subtract 5 from 30 to find Team F's score. Knowing the scores, you can now fill in the bar graph to show the data.

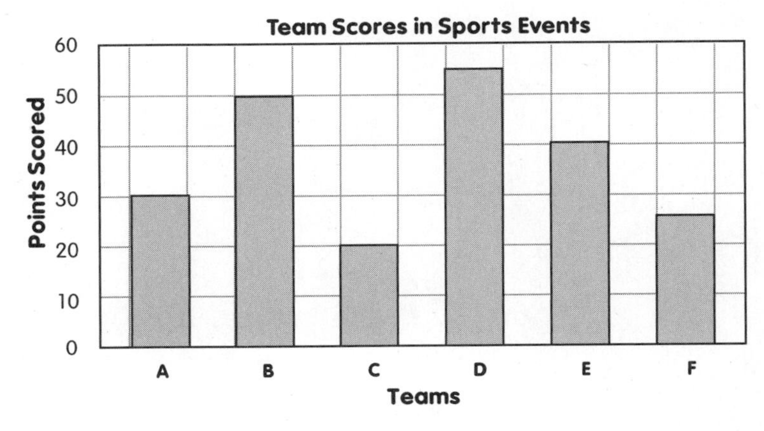

Tip: How can you check your answers?

You can reread the problem, referring to your bar graph to make sure it shows the correct information.

3.35 Selects appropriate operation (addition, subtraction, or multiplication) for a given problem situation.

3.36 Employs problem-solving strategies (e.g., draw a picture; make a chart, graph, or table; guess and check; look for a pattern).

3.39 Organizes data into charts and tables and constructs bar graphs using scales of one, two, five, or ten units and pictographs using scales of one, two, three, four, five, or ten units.

26. Answer: See discussion below.

Discussion

Tip: What symbol can you use to stand for less than 2 books?

Books Read by Third Graders	
October	
November	
December	
January	

Key: = 2 books   = 1 book

You can use half a square to stand for less than 2 books. Other possible observations: In October, the third graders read 2 fewer books than they did in January. In December, they read 6 more books than they did in November. In October and November combined, they read the same amount as they did in December. In January, they read 3 fewer books than they did in December.

Possible observations:

1. November and January had an odd number of books.

2. The third graders read 30 books all together.

3. In November, the class read the least amount of books.

Tip: How can you check your answer?

You can count the squares in the pictograph by 2's, and count the half-squares as 1's, comparing each total to the tally marks in the table.

3.28 Compares and orders whole numbers through 9,999.

3.39 Organizes data into charts and tables and constructs bar graphs using scales of one, two, five, or ten units and pictographs using scales of one, two, three, four, five, or ten units.

Answer Keys

Georgia Quality Core Curriculum Objectives

3.5 Recognizes properties (such as sides and angles) of geometric shapes (such as triangles, rectangles, circles, squares, and closed and not closed figures) and recognizes and names solid figures (such as cylinders, cones, spheres, and cubes).

3.6 Identifies and distinguishes among points, lines, line segments, rays, and angles.

3.7 Visualizes, draws, and compares geometric shapes in various positions/orientations.

3.8 Given a shape with a piece missing, selects the shape of the piece needed to complete the given shape.

3.9 Identifies geometric relations (parallel, inside, outside, same size, same shape, same size and shape, shorter/shortest, longer/longest, smaller/smallest, larger/largest), geometric transformations (same size and shape, but different position) and line of symmetry.

3.10 Sorts geometric shapes according to same shape (similar) and according to same shape and size (congruent).

3.33 Organizes elements of sets according to characteristics such as shading, color, shape, size, design, use, and number of sides.

3.38 Locates points on a map or grid.

1. Answer: A only straight lines

Discussion

Tip: What key words could you look for to give you clues?

Students should be familiar with a pentagon, but they may not have analyzed how the sides look. Ask these questions to help them focus on the sides and how they are formed.

Describe each side.
Is the side straight? How do you know?
Is the side curved? How do you know?

Have students draw pictures to show the difference between a straight and a curved line.

Encourage students to share their drawings with the class and explain the differences between curved and straight lines.

Item Numbers	Georgia QCC Objectives
1. A	3.6
2. G	3.5
3. D	3.5
4. G	3.5
5. C	3.5
6. J	3.6
7. C	3.6
8. H	3.38
9. A	3.38
10. F	3.7
11. C	3.7
12. G	3.33
13. B	3.10
14. G	3.7
15. C	3.9
16. J	3.8
17. C	3.9
18. G	3.10

19. Answers: A. rectangular prism, B. sphere, C. cone, D. cylinder

Discussion

Tip: How can picturing the objects in your mind help you solve the problem?

By picturing the objects in your mind, or by drawing them, you can compare and match them to the geometric shapes. Some kinds of cereal come in cylinders, but since the roll of pennies can only be a cylinder, the cereal box has to be a rectangular prism.

3.5 Recognizes properties (such as sides and angles) of geometric shapes (such as triangles, rectangles, circles, squares, and closed and not closed figures) and recognizes and names solid figures (such as cylinders, cones, spheres, and cubes).

3.34 Identifies information needed to solve a given problem.

20. Answer: See discussion below.

Discussion

Tip: How can drawing a picture help you solve this problem?

You can draw a triangle, a rectangle, and a square and count the sides and corners of each shape. Two triangles, two rectangles, or two squares can be different in size. The corners of a triangle can be different sizes. A rectangle can have different sizes in each pair of sides, but the corners are always the same. A square can be any size, but all sides and corners must be the same.

Shape	Number of Sides	Number of Corners
Triangle	3	3
Rectangle	4	4
Square	4	4

3.5 Recognizes properties (such as sides and angles) of geometric shapes (such as triangles, rectangles, circles, squares, and closed and not closed figures) and recognizes and names solid figures (such as cylinders, cones, spheres, and cubes).

3.9 Identifies geometric relations (parallel, inside, outside, same size, same shape, same size and shape, shorter/shortest, longer/longest, smaller/smallest, larger/largest), geometric transformations (same size and shape, but different position) and line of symmetry.

3.10 Sorts geometric shapes according to same shape (similar) and according to same shape and size (congruent).

21. Answer: 10 stars

Discussion

Tip: How can drawing a picture help you solve the problem?

Draw an X to represent the star on top. If each row has one less star than the one below, then each row has one more star than the one above. So you know that row two had 2 stars, row three had 3 stars, and row four had 4 stars.

X
XX
XXX
XXXX

Write a number sentence. $1 + 2 + 3 + 4 = 10$

3.31 Determines a pair of numbers or the missing element of a pair when given a relation or rule. Determines the relation or rule when given pairs of numbers.

3.35 Selects appropriate operation (addition, subtraction, or multiplication) for a given problem situation.

3.50 Applies mental computation strategies (such as counting up, counting back, simple compatible numbers, doubles, making ten, multiples of ten) to addition and subtraction, and to simple multiplication and division.

22. Answer: See discussion below.

Discussion

Tip: How can you find the pattern that repeats?

The pattern that has been established is circle, large triangle, small triangle, small triangle. To continue the pattern, the next two shapes should be circle, large triangle.

3.5 Recognizes properties (such as sides and angles) of geometric shapes (such as triangles, rectangles, circles, squares, and closed and not closed figures) and recognizes and names solid figures (such as cylinders, cones, spheres, and cubes).

3.31 Determines a pair of numbers or the missing element of a pair when given a relation or rule. Determines the relation or rule when given pairs of numbers.

3.36 Employs problem-solving strategies (e.g., draw a picture; make a chart, graph, or table; guess and check; look for a pattern).

23. Answer: Possible answers are shown.

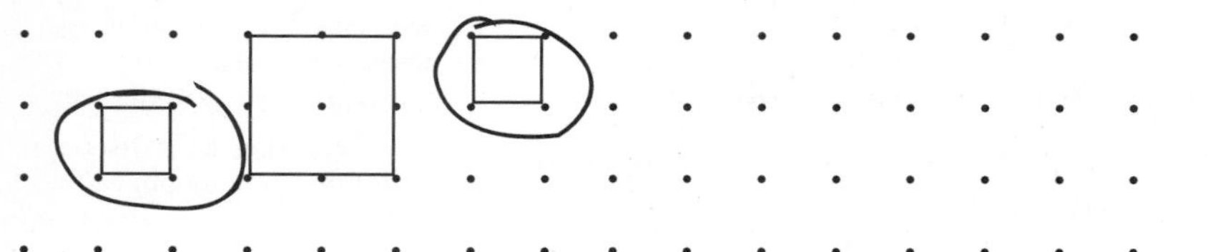

Discussion

Tip: What math word do you need to know? How can the dot paper help you draw figures that are the same size and shape?

If you draw squares whose sides cover the same amount of dots, you know that they are the same size. The square whose sides cover different amounts of dots is not congruent to the other two because it is a different size.

3.5 Recognizes properties (such as sides and angles) of geometric shapes (such as triangles, rectangles, circles, squares, and closed and not closed figures) and recognizes and names solid figures (such as cylinders, cones, spheres, and cubes).

3.10 Sorts geometric shapes according to same shape (similar) and according to same shape and size (congruent).

24. Answer: Pairs 1, 2, and 3 are congruent. Pair 4 is not.

Discussion

Tip: What is true of congruent figures?

To be congruent, two figures must be the same size and shape. If you can cut out one of the figures and place it exactly on top of the other one, then they are congruent. Flipping and turning them does not change the shape and size.

3.10 Sorts geometric shapes according to same shape (similar) and according to same shape and size (congruent).

3.36 Employs problem-solving strategies (e.g., draw a picture; make a chart, graph, or table; guess and check; look for a pattern).

25. Answer: 11 motorcycles

Discussion

Tip: In your picture, what do you need to show first? What important detail do you know that will help you solve the problem?

You can begin with an X to represent Ray's motorcycle. Then draw five X's to show the motorcycles to the right of Ray's motorcycle. If Ray's is in the middle, then there must be five motorcycles to the left of his. You can draw those five X's, and then count the X's to find that there are 11 in all.

Another strategy is to multiply the number of motorcycles on the right by 2 (to account for the ones on the left) and add 1 (Ray's) to that total.

(2 x 5) + 1 = 11 motorcycles

3.10 Sorts geometric shapes according to same shape (similar) and according to same shape and size (congruent).

3.35 Selects appropriate operation (addition, subtraction, or multiplication) for a given problem situation.

3.50 Applies mental computation strategies (such as counting up, counting back, simple compatible numbers, doubles, making ten, multiples of ten) to addition and subtraction, and to simple multiplication and division.

26. Answer: See discussion below.

Discussion

Tip: If you held a mirror up to each unfinished edge, what would you see?

The mirror would reflect the symmetrical other half of the figure.

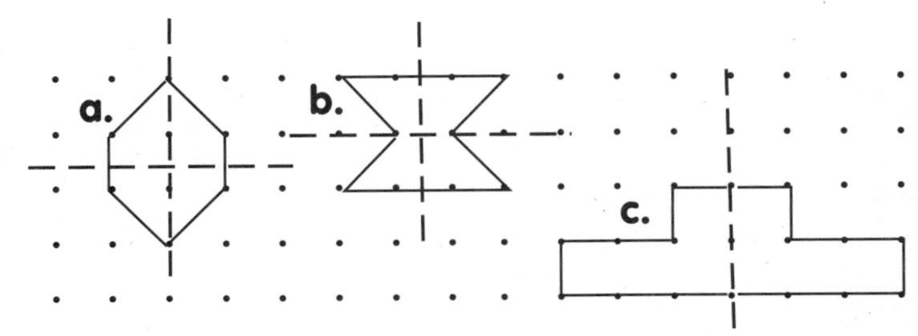

3.9 Identifies geometric relations (parallel, inside, outside, same size, same shape, same size and shape, shorter/shortest, longer/longest, smaller/smallest, larger/largest), geometric transformations (same size and shape, but different position) and line of symmetry.

3.10 Sorts geometric shapes according to same shape (similar) and according to same shape and size (congruent).

3.36 Employs problem-solving strategies (e.g., draw a picture; make a chart, graph, or table; guess and check; look for a pattern).

27. Answer: 1. B; 2. C; 3. B; 4. B; 5. A.

Discussion

Tip: What do you know about rectangles, squares, and triangles that will help you?

You need to recall the definitions of square, rectangle, and triangle before you can solve this problem. Knowing that a triangle has 3 sides helps you to pick out sets 1, 3, and 4 for triangles, since those sets have 3 sticks each. Comparing the sizes of the sticks helps you decide whether each triangle will have 0 equal sides, 2 equal sides, or 3 equal sides.

You know that a square has 4 equal sides. So you can pick out set 2, since all those sticks are the same size.

Set 5 will make a rectangle, since it has 4 sticks (4 sides) of 2 different sizes.

Tip: How can making a model with sticks help you solve the problem?

You can use toothpicks or strips of paper to make a model for each set of sticks.

3.5 Recognizes properties (such as sides and angles) of geometric shapes (such as triangles, rectangles, circles, squares, and closed and not closed figures) and recognizes and names solid figures (such as cylinders, cones, spheres, and cubes).

3.16 Estimates or predicts measures of length, weight, volume, capacity, and temperature.

3.36 Employs problem-solving strategies (e.g., draw a picture; make a chart, graph, or table; guess and check; look for a pattern).

28. Answer: See discussion below.

Discussion

Tip: What is alike about the shapes? What is different about them?

The shapes are similar in that both have four sides and four equal corners. The only difference is that in a square, the sides are the same length.

Shape A is both a square and a rectangle, and shape B is a rectangle. Both of them have four corners that are equal. Both of them have four sides. In a square, all the sides are the same length. In a rectangle, opposite sides are the same length.

Tip: How can you check that your explanation is clear and complete?

Don't forget to mention the number of sides, the size of the corners, and the length of the sides.

3.5 Recognizes properties (such as sides and angles) of geometric shapes (such as triangles, rectangles, circles, squares, and closed and not closed figures) and recognizes and names solid figures (such as cylinders, cones, spheres, and cubes).

3.9 Identifies geometric relations (parallel, inside, outside, same size, same shape, same size and shape, shorter/shortest, longer/longest, smaller/smallest, larger/largest), geometric transformations (same size and shape, but different position) and line of symmetry.

3.10 Sorts geometric shapes according to same shape (similar) and according to same shape and size (congruent).

Answer Keys

Georgia Quality Core Curriculum Objectives

3.2 Relates models separated into 10 equivalent parts to the language of decimals, such as "five-tenths" rather than "point five".

3.3 Identifies and writes fractions to describe parts of a whole using both regions and discrete sets (halves, thirds, fourths, sixths, eighths, and tenths).

3.4 Compares fractions with like denominators and explores comparison of fractions with unlike denominators using models.

3.36 Employs problem-solving strategies (e.g., draw a picture; make a chart, graph, or table; guess and check; look for a pattern).

3.37 Solves one- and two-step word problems related to appropriate third grade objectives. Includes oral and written problems and problem with extraneous information as well as information from sources such as pictographs, bar graphs, tables, and charts.

1. Answer: B 6 parts

Discussion

Tip: Count the total number of parts in the fraction model.

Students should recognize from the model that the parts are equal and that all of the parts make up the whole, not just the shaded part. Ask questions such as these to help students understand the meaning of a written fraction:

Is each part of the fraction model equal?
Why is it important to know how many parts make up the whole when you are writing a fraction?
How do you use the shaded part of the model when writing a fraction?

Also explain what each part of the fraction corresponds to—the numerator corresponds to the shaded part and the denominator corresponds to the number of parts that make up the whole.

Item Numbers	Georgia QCC Objectives
1. B	3.3
2. F	3.3
3. D	3.3
4. G	3.3
5. B	3.36
6. F	3.4
7. B	3.36, 3.37
8. G	3.36, 3.37
9. B	3.36
10. G	3.2
11. A	3.4
12. H	3.36
13. B	
14. G	
15. A	
16. F	
17. B	3.36
18. G	3.36
19. A	3.36

20. Answer: Joe is working on $\frac{3}{8}$ of the wall.

Discussion

Tip: How many parts are there in all?

The wall has been divided into 8 parts. Joe is responsible for 3 of those parts. So Joe is working on $\frac{3}{8}$ of the wall.

3.3 Identifies and writes fractions to describe parts of a whole using both regions and discrete sets (halves, thirds, fourths, sixths, eighths, and tenths).

3.36 Employs problem-solving strategies (e.g., draw a picture; make a chart, graph, or table; guess and check; look for a pattern).

21. Answer: blue beads: $\frac{2}{5}$ or two fifths

Discussion

Tip: How many equal sections does the bead case have? How many kinds of blue beads does Cassie have?

The bead case has 5 sections. Of these, 2 sections are for blue beads, or $\frac{2}{5}$ of the bead case.

3.3 Identifies and writes fractions to describe parts of a whole using both regions and discrete sets (halves, thirds, fourths, sixths, eighths, and tenths).

3.36 Employs problem-solving strategies (e.g., draw a picture; make a chart, graph, or table; guess and check; look for a pattern).

22. Answer: See discussion below.

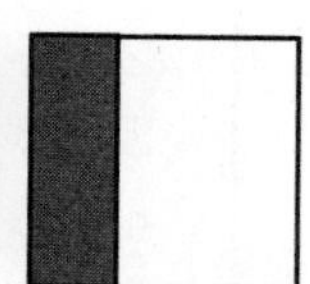
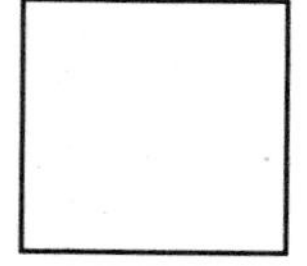

Discussion

Tip: How many sections are completely painted? How many are partly painted?

Three sections are completely painted, so you can shade in the first 3 squares. Only one third of the next section is painted, so you can shade in one third of it. Leave the fifth square blank.

3.3 Identifies and writes fractions to describe parts of a whole using both regions and discrete sets (halves, thirds, fourths, sixths, eighths, and tenths).

3.36 Employs problem-solving strategies (e.g., draw a picture; make a chart, graph, or table; guess and check; look for a pattern).

23. Answer: 5 mini-pizzas

Discussion

Tip: How can drawing a picture help you solve this problem?

Possible solution:

Each person gets

So draw enough whole pizzas for four people:

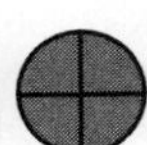
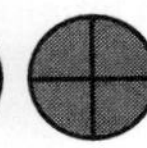

Then color one-fourth for each person. 

That equals five pizzas.

3.3 Identifies and writes fractions to describe parts of a whole using both regions and discrete sets (halves, thirds, fourths, sixths, eighths, and tenths).

3.36 Employs problem-solving strategies (e.g., draw a picture; make a chart, graph, or table; guess and check; look for a pattern).

24. Answer: Fiona

Discussion

Tip: How can drawing a picture for each fraction help you decide?

You can use a fraction bar to compare $\frac{2}{5}$ and $\frac{2}{3}$. You will see that $\frac{2}{3}$ is a larger amount than $\frac{2}{5}$. When comparing fractions, remember that the larger the denominator, the smaller the pieces. An easy way to remember this is to picture $\frac{1}{10}$ and $\frac{1}{5}$. You know that $\frac{1}{10}$ is much smaller than $\frac{1}{5}$. In the same way, $\frac{1}{5}$ is smaller than $\frac{1}{3}$, so $\frac{2}{5}$ is smaller than $\frac{2}{3}$. If the numerators are the same, then the fraction with the smaller denominator will be a bigger piece.

3.4 Compares fractions with like denominators and explores comparison of fractions with unlike denominators using models.

3.36 Employs problem-solving strategies (e.g., draw a picture; make a chart, graph, or table; guess and check; look for a pattern).

25. Answer: $\frac{3}{7}$ of the plates are red.

Discussion

Tip: How can drawing a picture help you see the whole group and its parts?

You can draw 7 plates to show the number in the group. Then you can color 4 plates to show the part that is blue. The 3 remaining plates are the part of the group that is red.

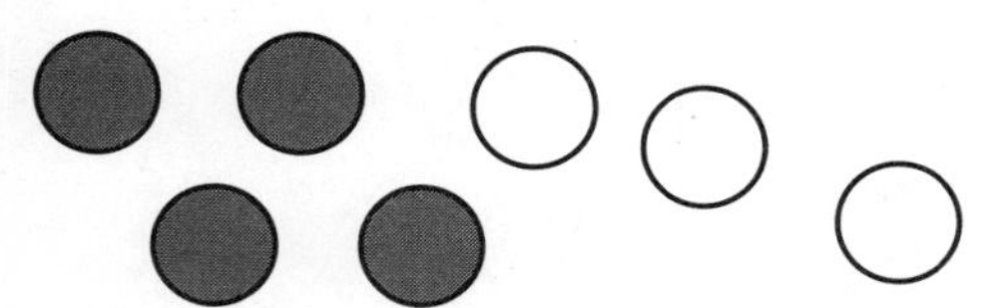

3.3 Identifies and writes fractions to describe parts of a whole using both regions and discrete sets (halves, thirds, fourths, sixths, eighths, and tenths).

3.33 Organizes elements of sets according to characteristics such as shading, color, shape, size, design, use, and number of sides.

3.35 Selects appropriate operation (addition, subtraction, or multiplication) for a given problem situation.

3.36 Employs problem-solving strategies (e.g., draw a picture; make a chart, graph, or table; guess and check; look for a pattern).

---

26. Answer: $12.50

Discussion

Tip: What is the pattern?

The pattern is $5 for every 100 papers. Add $2.50 (half of $5) for every 50 papers (half of 100 papers). So for 250 papers, Jake would get $12.50.

Another strategy is to make a list, and notice how each amount goes up.

100	$5.00
150	$7.50
200	$10.00
250	

Every time 50 papers is added in column 1, $2.50 is added in column 2.

3.31 Determines a pair of numbers or the missing element of a pair when given a relation or rule. Determines the relation or rule when given pairs of numbers.

3.36 Employs problem-solving strategies (e.g., draw a picture; make a chart, graph, or table; guess and check; look for a pattern).

3.50 Applies mental computation strategies (such as counting up, counting back, simple compatible numbers, doubles, making ten, multiples of ten) to addition and subtraction, and to simple multiplication and division.

---

27. Answer: 2.3

Discussion

Tip: What is a mixed decimal?

A mixed decimal is a number that is made up of a whole number and a decimal. The model shows that 2 areas of the parking lot are filled. The third area has 10 spaces, but only 3 are filled. In other words, three tenths of the third area is filled. $2 + \frac{3}{10}$ is the same as 2 + 0.3, or 2.3.

3.35 Selects appropriate operation (addition, subtraction, or multiplication) for a given problem situation.

3.36 Employs problem-solving strategies (e.g., draw a picture; make a chart, graph, or table; guess and check; look for a pattern).

3.50 Applies mental computation strategies (such as counting up, counting back, simple compatible numbers, doubles, making ten, multiples of ten) to addition and subtraction, and to simple multiplication and division.

28. Answer: George could be right if he had half of a large pizza and Corinne had half of a small pizza.

Discussion

Tip: When is half of one thing more than half of another?

Half of a large pizza is bigger than half of a small pizza. Even though both ate half of a pizza, George could have eaten more than Corinne if his pizza was larger. In order for Corinne to be right, both pizzas would have to be the same size. In order for George to be right, his pizza would have to be larger.

Tip: Be sure that your explanation is clear and complete.

3.3 Identifies and writes fractions to describe parts of a whole using both regions and discrete sets (halves, thirds, fourths, sixths, eighths, and tenths).

3.4 Compares fractions with like denominators and explores comparison of fractions with unlike denominators using models.

3.35 Selects appropriate operation (addition, subtraction, or multiplication) for a given problem situation.

---

29. Answer: They have planted $\frac{9}{12}$ (or $\frac{3}{4}$) of the rows.

Discussion

Tip: What do you need to find out first?

You need to find out the total number of rows they have planted. Add 3 (Ben's rows) and 6 (his mother's rows) to get 9. They have planted 9 out of 12 rows. This can be expressed as $\frac{9}{12}$ or $\frac{3}{4}$.

Tip: How can you check your answer?

You can draw a picture in which each row is represented by an bar. Draw 12 bars. Circle 3 bars and label them "Ben." Circle 6 rows and label them "his mother." You can see that $\frac{9}{12}$ of the bars have been circled. This also makes $\frac{3}{4}$ of the bars.

3.3 Identifies and writes fractions to describe parts of a whole using both regions and discrete sets (halves, thirds, fourths, sixths, eighths, and tenths).

3.35 Selects appropriate operation (addition, subtraction, or multiplication) for a given problem situation.

3.50 Applies mental computation strategies (such as counting up, counting back, simple compatible numbers, doubles, making ten, multiples of ten) to addition and subtraction, and to simple multiplication and division.

Answer Keys

Georgia Quality Core Curriculum Objectives

3.11 Measures, using appropriate instruments and appropriate units, length, capacity, weight/mass, time, and temperature. Length: millimeter, inch, centimeter, foot, meter, yard, kilometer, mile. Capacity: milliliter, ounce, liter, cup, pint (liquid and dry), quart (liquid and dry), gallon. Weight/Mass: gram, ounce, kilogram, pound. Time: second, week, minute, month, hour, year, day, decade, century. Temperature: degree Fahrenheit, degree Celsius.

3.12 Measures using concrete materials such as string to find perimeter and circumference; squares or tiles to find area; and cubes to find volume. Determines perimeter by adding lengths of sides.

3.13 Selects appropriate customary and metric units of measure. Length: millimeter, inch, centimeter, foot, meter, yard, kilometer, mile. Capacity: milliliter, ounce, liter, cup, pint (liquid and dry), quart (liquid and dry), gallon. Weight/Mass: gram, ounce, kilogram, pound. Time: second, week, minute, month, hour, year, day, decade, century. Temperature: degree Fahrenheit, degree Celsius.

3.16 Estimates or predicts measures of length, weight, volume, capacity, and temperature.

3.34 Identifies information needed to solve a given problem.

3.36 Employs problem-solving strategies (e.g., draw a picture; make a chart, graph, or table; guess and check; look for a pattern).

3.37 Solves one- and two-step word problems related to appropriate third grade objectives. Includes oral and written problems and problem with extraneous information as well as information from sources such as pictographs, bar graphs, tables, and charts.

1. Answer: A inches

Discussion

Tip: Which units of measurement would not be good choices?

Students should be familiar with the common tools for measuring customary units, such as a ruler and a yardstick. Ask questions to encourage them to think about the best unit of measure:

> Which tools could you choose to measure how wide a backpack is?
> What do you estimate your measurement might be?
> Explain why the other answer choices would not be the best unit of measure.

You could bring in a backpack and other common items. Have students estimate and then measure these items with a ruler and/or yardstick.

Item Numbers	Georgia QCC Objectives
1. A	3.13
2. G	3.13
3. C	3.11
4. H	3.11
5. C	3.13
6. F	3.16
7. D	3.16
8. F	3.13
9. A	3.16
10. G	3.13
11. B	3.11
12. J	3.11
13. D	3.12
14. J	3.12
15. D	3.12
16. H	3.12, 3.37
17. B	3.12
18. H	3.12, 3.34, 3.36, 3.37
19. C	3.12, 3.37

20. Answer: 68 miles

Discussion

Tip: What information do you need to solve the problem?

The only information you need is the total number of miles each person drove for the round trip. The number of hours each person stayed is not needed. To find the total number of miles each person drove, double the number of one-way miles. If Patrick drove 34 miles one way, then he drove 68 miles total. To find this, you can write a number sentence: 34 miles + 34 miles = 68 miles. If Natalie drove 68 miles one way, then she drove 136 miles total. To find out which person drove more, you are looking for a difference. This means that you subtract. You can write a number sentence.

136 miles – 68 miles = 68 miles.

3.34 Identifies information needed to solve a given problem.

3.50 Applies mental computation strategies (such as counting up, counting back, simple compatible numbers, doubles, making ten, multiples of ten) to addition and subtraction, and to simple multiplication and division.

21. Answer: Basking, Great White, Thresher, Mako, and Horn

Discussion

Tip: What are you trying to find out? Restate the problem in your own words.

You can list the lengths of the sharks in order from the longest to the shortest: 25 feet, 18 feet, 5 feet, 9 feet, and 4 feet. When you have ordered the lengths, label each one with the name of the shark. Then you can see the order in which they should appear on the poster.

Another strategy is to locate each length on a number line, label the number with the name of the shark, and then read backward from longest to shortest.

3.16 Estimates or predicts measures of length, weight, volume, capacity, and temperature.

3.28 Compares and orders whole numbers through 9,999.

3.36 Employs problem-solving strategies (e.g., draw a picture; make a chart, graph, or table; guess and check; look for a pattern).

22. Answer: 4 quarts

Discussion

Tip: How many cups are in a quart?

Megan's mother needs 16 cups of juice. Since there are 4 cups in a quart, you can divide 16 by 4 to find out how many quarts are needed. $16 \div 4 = 4$.

Another strategy you can use is guess and check. A guess of 3 quarts (or 12 cups) would not be enough. A guess of 4 quarts (or 16 cups) is correct.

3.16 Estimates or predicts measures of length, weight, volume, capacity, and temperature.

3.35 Selects appropriate operation (addition, subtraction, or multiplication) for a given problem situation.

3.36 Employs problem-solving strategies (e.g., draw a picture; make a chart, graph, or table; guess and check; look for a pattern).

3.50 Applies mental computation strategies (such as counting up, counting back, simple compatible numbers, doubles, making ten, multiples of ten) to addition and subtraction, and to simple multiplication and division.

23. Answer: See discussion below.

Discussion

Tip: When it's cooler, is the temperature higher or lower?

Cooler temperatures are lower. Since the first thermometer reads 85 degrees, the second one should read 75 degrees.

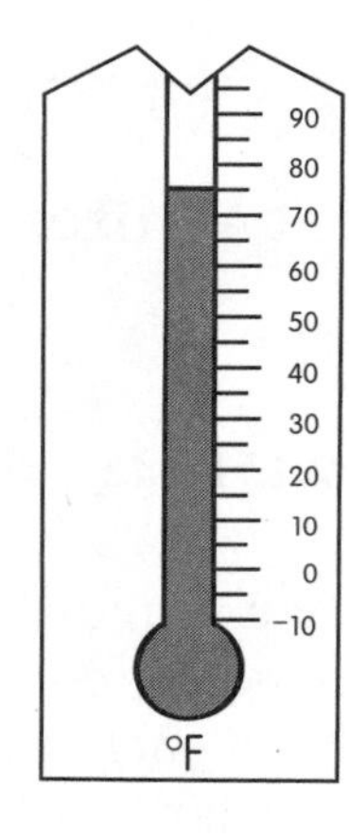

3.11 Measures, using appropriate instruments and appropriate units, length, capacity, weight/mass, time, and temperature. Length: millimeter, inch, centimeter, foot, meter, yard, kilometer, mile. Capacity: milliliter, ounce, liter, cup, pint (liquid and dry), quart (liquid and dry), gallon. Weight/Mass: gram, ounce, kilogram, pound. Time: second, week, minute, month, hour, year, day, decade, century. Temperature: degree Fahrenheit, degree Celsius.

3.35 Selects appropriate operation (addition, subtraction, or multiplication) for a given problem situation.

3.50 Applies mental computation strategies (such as counting up, counting back, simple compatible numbers, doubles, making ten, multiples of ten) to addition and subtraction, and to simple multiplication and division.

24. Answer: Accept responses between 50°F and 70°F.

Discussion

Tip: What was the highest temperature of the week? What was the lowest?

The highest temperature was 70°F, and the lowest was about 55°F. There may be a warming trend, so the temperature might go up some more to about 65°F to 70°F.

3.28 Compares and orders whole numbers through 9,999.

3.36 Employs problem-solving strategies (e.g., draw a picture; make a chart, graph, or table; guess and check; look for a pattern).

3.40 Collects, reads, interprets, and compares data in charts, tables, and graphs.

25. Answer: 44 feet

Discussion

Tip: How can drawing a picture help you solve the problem?

You need to find out the perimeter of the room. The perimeter is the distance around the room. You can add the length of each side.

12 + 10 + 12 + 10 = 44 feet

Another strategy is to multiply each side by 2 and add the two products.

(12 x 2) + (10 x 2) = 44 feet

3.12 Measures using concrete materials such as string to find perimeter and circumference; squares or tiles to find area; and cubes to find volume. Determines perimeter by adding lengths of sides.

3.35 Selects appropriate operation (addition, subtraction, or multiplication) for a given problem situation.

3.36 Employs problem-solving strategies (e.g., draw a picture; make a chart, graph, or table; guess and check; look for a pattern).

3.50 Applies mental computation strategies (such as counting up, counting back, simple compatible numbers, doubles, making ten, multiples of ten) to addition and subtraction, and to simple multiplication and division.

26. Answer: 160 ft

Discussion

Tip: How can you use the diagram to solve the problem?

The diagram shows the length of each side of the playground. You can add all the measures to find the perimeter.

Start at one corner. Write each measure as you point to each one.

30 ft + 50 ft + 20 ft + 29 ft + 10 ft + 30 ft = 160 ft

3.12 Measures using concrete materials such as string to find perimeter and circumference; squares or tiles to find area; and cubes to find volume. Determines perimeter by adding lengths of sides.

3.35 Selects appropriate operation (addition, subtraction, or multiplication) for a given problem situation.

3.36 Employs problem-solving strategies (e.g., draw a picture; make a chart, graph, or table; guess and check; look for a pattern).

27. Answer: 6:00

Discussion

Tip: What time does Randi get to the library? When does she leave?

You can trace Randi's route on the map, marking the times she arrives and leaves each place.

Another strategy is to add the total walking time (1 hour) to library time ($1\frac{1}{2}$ hour) to music store time (1 hour) and count forward $2\frac{1}{2}$ hours from 3:30.

Tip: How can you check your answer?

You can act it out on a clock with movable hands.

3.14 Determines and estimates amounts of money up to $5.00. Includes amounts spent, change received, and equivalent amounts.

3.34 Identifies information needed to solve a given problem.

3.36 Employs problem-solving strategies (e.g., draw a picture; make a chart, graph, or table; guess and check; look for a pattern).

28. Possible answer: See chart, and discussion below.

Discussion

Tip: What facts on the chart will help you solve the problem?

You can use the weight of the pet carrier and the weight of Muffin in the pet carrier.

You can subtract the weight of the pet carrier from the weight of Muffin in the pet carrier to find how much Muffin weighs.

30 pounds – 5 pounds = 25 pounds

Once you know how much Muffin weighs, you can add his weight to Marianne's to complete the chart.

65 pounds + 25 pounds = 90 pounds

Tip: How can you check that your answers are correct?
How can you check that your explanations are clear?

You can check your subtraction by adding. You can check your addition by subtracting.

3.16 Estimates or predicts measures of length, weight, volume, capacity, and temperature.

3.35 Selects appropriate operation (addition, subtraction, or multiplication) for a given problem situation.

3.40 Collects, reads, interprets, and compares data in charts, tables, and graphs.

29. Possible answers: 6 cm by 3 cm, 5 cm by 4 cm

Discussion

Tip: How can using the strategy Guess and Check help you solve the problem?

You can guess a length for the longer sides of one rectangle and see if it will work. For example, if one side is 6 cm, then you know the opposite side must be 6 cm. Two sides of 6 cm each will use 12 cm of ribbon, leaving 6 cm, or 3 cm each, for the other two sides. You can follow the same procedure using 5, 7, and 8 for the longer sides.

The strategy Guess and Check will lead you to discover that 9 cm is too long for any pair of sides, as this would use all the ribbon for just two sides.

You know that both sides are rectangles because they have 4 sides and 4 equal corners, with opposite sides being the same length.

Tip: How can you check your answer?

You can add the lengths of the sides of your rectangles to make sure they add up to 18 cm.

3.5 Recognizes properties (such as sides and angles) of geometric shapes (such as triangles, rectangles, circles, squares, and closed and not closed figures) and recognizes and names solid figures (such as cylinders, cones, spheres, and cubes).

3.12 Measures using concrete materials such as string to find perimeter and circumference; squares or tiles to find area; and cubes to find volume. Determines perimeter by adding lengths of sides.

3.16 Estimates or predicts measures of length, weight, volume, capacity, and temperature.

3.35 Selects appropriate operation (addition, subtraction, or multiplication) for a given problem situation.

3.36 Employs problem-solving strategies (e.g., draw a picture; make a chart, graph, or table; guess and check; look for a pattern).

Answer Keys

Georgia Quality Core Curriculum Objectives

3.12 Measures using concrete materials such as string to find perimeter and circumference; squares or tiles to find area; and cubes to find volume. Determines perimeter by adding lengths of sides.

3.35 Selects appropriate operation (addition, subtraction, or multiplication) for a given problem situation.

3.36 Employs problem-solving strategies (e.g., draw a picture; make a chart, graph, or table; guess and check; look for a pattern).

3.37 Solves one- and two-step word problems related to appropriate third grade objectives. Includes oral and written problems and problem with extraneous information as well as information from sources such as pictographs, bar graphs, tables, and charts.

3.42 Multiples whole numbers up to two-digit by one-digit numbers using models and three-digit by one-digit numbers using computational strategies.

3.44 Relates division to multiplication and uses models such as partitioning, and repeated subtraction to divide one- and two-digit numbers by one-digit numbers without and with remainders.

3.47 Selects appropriate symbol (+, –, x, —, <, >, =) for use in a number sentence.

3.49 Recalls basic multiplication facts through 9 x 9.

3.50 Applies mental computation strategies (such as counting up, counting back, simple compatible numbers, doubles, making ten, multiples of ten) to addition and subtraction, and to simple multiplication and division.

1. Answer: C 192

Discussion

Tip: How many rows of base-ten blocks would you need? How many blocks would be in each row?

To reinforce how to use base-ten blocks for multiplication, write this expression on the board: 2×41. Use overhead base-ten blocks to show the arrangement as the discussion progresses.

How many rows of blocks do I need?
How do you know?
How many blocks would be in each row?
How do you know?
How do I find the product using the blocks I have?

Make sure that in the discussion students tell how many ones and tens there are. They should also tell how you write tens as a number. For example, 4 tens is 40.

Item Numbers	Georgia QCC Objectives
1. C	3.42, 3.49
2. J	3.42, 3.49
3. B	3.42, 3.49
4. G	3.42, 3.49
5. A	3.35, 3.37, 3.49
6. G	3.44
7. A	3.44
8. J	3.44
9. C	3.44
10. G	3.35, 3.37
11. D	3.35, 3.36, 3.37 3.47
12. F	3.35, 3.36, 3.37 3.47
13. B	3.42
14. J	3.42
15. B	3.35, 3.37
16. H	3.35, 3.37, 3.49
17. A	3.12, 3.36
18. F	3.50
19. D	3.50

20. Answer: 73 subscriptions

Discussion

Tip: How many of each base-ten block can you use to model the problem?

You can use four base-ten blocks and three ones to model the first 43 subscriptions. Then you can add three more base-ten blocks to model the 10 subscriptions brought in by the next three students. Counting the blocks is one way to find the answer.

Another strategy is to begin with 43 and then count on by tens 3 times to get to 73.

3.22 Identifies place value through thousands and identifies the number of thousands, hundreds, tens, and ones in a given number.

3.35 Selects appropriate operation (addition, subtraction, or multiplication) for a given problem situation.

3.50 Applies mental computation strategies (such as counting up, counting back, simple compatible numbers, doubles, making ten, multiples of ten) to addition and subtraction, and to simple multiplication and division.

21. Answer: Yes, he has enough. He has $3.05 left over, and he only needs $2.50.

Discussion

Tip: What do you need to figure out first?

You need to know how much Larry has left over after buying the baseball cards. Since you don't need an exact number (the question asks only if he has enough, not exactly how much he has left), you can estimate by rounding $4.95 up to $5.00 and subtracting $5.00 from $8.00. Now you know that Larry has about $3.00 left. The next thing you need to find out is how much 5 game tokens will cost. You can add 50 cents 5 times, or multiply 50 cents by 5 to get $2.50. Larry needs $2.50, and he has about $3.00, so he has enough.

Another strategy is to use play money to act it out.

3.35 Selects appropriate operation (addition, subtraction, or multiplication) for a given problem situation.

3.36 Employs problem-solving strategies (e.g., draw a picture; make a chart, graph, or table; guess and check; look for a pattern).

3.50 Applies mental computation strategies (such as counting up, counting back, simple compatible numbers, doubles, making ten, multiples of ten) to addition and subtraction, and to simple multiplication and division.

22. Possible answers: triangle with 20-in. sides, square with 15-in. sides

Discussion

Tip: Can there be more than one answer?

Yes, your figure may have any number of equal sides. You can divide the perimeter by the number of sides to find the length of each side:

perimeter ÷ number of sides = length of each side

$60 \div 3 = 20$
$60 \div 4 = 15$
$60 \div 5 = 12$
$60 \div 6 = 10$

3.5 Recognizes properties (such as sides and angles) of geometric shapes (such as triangles, rectangles, circles, squares, and closed and not closed figures) and recognizes and names solid figures (such as cylinders, cones, spheres, and cubes).

3.12 Measures using concrete materials such as string to find perimeter and circumference; squares or tiles to find area; and cubes to find volume. Determines perimeter by adding lengths of sides.

3.36 Employs problem-solving strategies (e.g., draw a picture; make a chart, graph, or table; guess and check; look for a pattern).

3.50 Applies mental computation strategies (such as counting up, counting back, simple compatible numbers, doubles, making ten, multiples of ten) to addition and subtraction, and to simple multiplication and division.

23. Answer: Hector is not right. Both have the same amount of shelf space.

Discussion

Tip: What do you need to find out first?

You need to find out how much shelf space each person has. Dina has 6 shelves of 3 feet each. You can multiply 6 by 3, or you can add 6 + 6 + 6 to find that she has 18 feet of shelf space. Hector's 24 inches of shelf space is the same as 2 feet. If he has 9 shelves that are 2 feet each, he has 18 feet of shelf space.

Tip: Be sure your answer is clear and complete.

Explain how you figured out the shelf space for each person. If you converted inches to feet, explain why you did so.

3.16 Estimates or predicts measures of length, weight, volume, capacity, and temperature.

3.28 Compares and orders whole numbers through 9,999.

3.35 Selects appropriate operation (addition, subtraction, or multiplication) for a given problem situation.

3.50 Applies mental computation strategies (such as counting up, counting back, simple compatible numbers, doubles, making ten, multiples of ten) to addition and subtraction, and to simple multiplication and division.

24. Answer: 48 more crocuses than daffodils

Discussion

Tip: What do you need to find out first?

You need to find out how many crocuses and daffodils Stella planted. Each bulb represents 12 bulbs, so you multiply the number of bulbs in the pictograph by 12 to find the total for each kind.

12 x 8 = 96 crocuses

12 x 4 = 48 daffodils

Then you can subtract to find the difference.

96 - 48 = 48 more crocuses

Another strategy is to find the difference in the number of pictures in the pictograph.

8 – 4 = 4 pictures

Then you can multiply the number of pictures by 12 to find the number of bulbs represented by the difference.

4 x 12 = 48 more crocuses

3.35 Selects appropriate operation (addition, subtraction, or multiplication) for a given problem situation.

3.36 Employs problem-solving strategies (e.g., draw a picture; make a chart, graph, or table; guess and check; look for a pattern).

3.40 Collects, reads, interprets, and compares data in charts, tables, and graphs.

3.50 Applies mental computation strategies (such as counting up, counting back, simple compatible numbers, doubles, making ten, multiples of ten) to addition and subtraction, and to simple multiplication and division.

25. Answer: 3,000 pounds

Discussion

Tip: How can using mental math help you solve the problem?

You can add 15 and 15 mentally, since when you add hundreds and thousands, only the hundreds and thousands places change value. The ones and tens places remain the same.

Another strategy is to write a number sentence.

1,500 pounds + 1,500 pounds = 3,000 pounds.

3.16 Estimates or predicts measures of length, weight, volume, capacity, and temperature.

3.36 Employs problem-solving strategies (e.g., draw a picture; make a chart, graph, or table; guess and check; look for a pattern).

3.50 Applies mental computation strategies (such as counting up, counting back, simple compatible numbers, doubles, making ten, multiples of ten) to addition and subtraction, and to simple multiplication and division.

26. Answer: about 80

Discussion

Tip: How many 15-minute periods are there in one hour? How can rounding help you solve the problem?

Since the question asks for an estimate, you can round 19 up to the nearest 10, which would be 20. Then, since there are four 15-minute periods in one hour, you can multiply 20 pieces of litter by 4.

$20 \times 4 = 80$

Another method is to add 20 four times.

$20 + 20 + 20 + 20 = 80$

3.15 Tells time to the minute and measures elapsed time, and measures time before and after the hour.

3.36 Employs problem-solving strategies (e.g., draw a picture; make a chart, graph, or table; guess and check; look for a pattern).

3.50 Applies mental computation strategies (such as counting up, counting back, simple compatible numbers, doubles, making ten, multiples of ten) to addition and subtraction, and to simple multiplication and division.

27. Answer: 70 cups

Discussion

Tip: What operation can you use to find the answer?

For each 5 cups, Julia got 1 free ticket. If she has 14 free tickets, then she bought 5 cups 14 times. You can multiply 5 cups of juice by 14 tickets: $5 \times 14 = 70$.

Another strategy is to count by 5's 14 times until you get to 70.

3.35 Selects appropriate operation (addition, subtraction, or multiplication) for a given problem situation.

3.36 Employs problem-solving strategies (e.g., draw a picture; make a chart, graph, or table; guess and check; look for a pattern).

3.50 Applies mental computation strategies (such as counting up, counting back, simple compatible numbers, doubles, making ten, multiples of ten) to addition and subtraction, and to simple multiplication and division.

28. Answer: 25 cards

Discussion

Tip: What information do you need to solve the problem?

You need to decide what information is needed and what information is not needed. The question is about the number of cards Maggie had left. Therefore, the information about the 3 pieces of bubble gum is not needed. To find out how many cards Maggie had left, you first must figure out how many she had to begin with. If one pack has 10, then 3 packs would have 30. You can either add 10 + 10 + 10, or you can multiply 10 x 3 to get 30. If she gave 5 of those 30 cards to her brother, she had 25 cards left.

Tip: How can making a model help you solve the problem?

You can use 30 counters to represent 3 packs of 10 cards each. Model giving away 5 counters to find that there are 25 left.

3.34 Identifies information needed to solve a given problem.

3.35 Selects appropriate operation (addition, subtraction, or multiplication) for a given problem situation.

3.36 Employs problem-solving strategies (e.g., draw a picture; make a chart, graph, or table; guess and check; look for a pattern).

3.50 Applies mental computation strategies (such as counting up, counting back, simple compatible numbers, doubles, making ten, multiples of ten) to addition and subtraction, and to simple multiplication and division.

29. Answer: 12 more brushes

Discussion

Tip: What do you need to figure out first?

The first thing you need to determine is that 4 is half of 8. If 4 jars hold 10 brushes each, that is a total of 40 brushes. The other 4 jars have 12 brushes each, for a total of 48 brushes. 40 brushes + 48 brushes = 88 brushes. You know you need 100 brushes. You can count on from 88 to 100 to find that 12 brushes are still needed. Another method is to subtract 88 from 100.

Another strategy is to draw a picture representing 8 jars, writing 10 on half of the jars and 12 on the other half. Then you can add the numbers and subtract the total from 100.

Tip: Be sure that your explanation is clear and complete.

Explain the steps you followed to arrive at the answer.

3.35 Selects appropriate operation (addition, subtraction, or multiplication) for a given problem situation.

3.36 Employs problem-solving strategies (e.g., draw a picture; make a chart, graph, or table; guess and check; look for a pattern).

3.50 Applies mental computation strategies (such as counting up, counting back, simple compatible numbers, doubles, making ten, multiples of ten) to addition and subtraction, and to simple multiplication and division.

Notes